Build Your Own
TEAMS OF ROBOTS
with LEGO® MINDSTORMS® NXT
and Bluetooth®

**Cameron Hughes, Tracey Hughes,
Trevor Watkins, and Bob Kramer**

New York   Chicago   San Francisco   Lisbon
London   Madrid   Mexico City   Milan   New Delhi
San Juan   Seoul   Singapore   Sydney   Toronto

Sponsoring Editor Roger Stewart	**Project Manager** Patricia Wallenburg, TypeWriting	**Indexer** Jack Lewis
Editing Supervisor Stephen M. Smith	**Copy Editor** James K. Madru	**Art Director, Cover** Jeff Weeks
Production Supervisor Pamela A. Pelton	**Proofreader** Paul Tyler	**Composition** TypeWriting
Acquisitions Coordinator Molly T. Wyand		

The authors would like to dedicate this book to those who helped us, showed patience, and gave us inspiration:

To Barbara, who gave Bob his first LEGO set in 1974, and to Betsy, who graciously gave Bob the precious time and space to work on the material and to build that amazing cube solver.

To Lael, who loaned us Trevor and his expertise for far too many long hours.

To Takeo, whose 60th birthday celebration made the robot spark our priority.

To Mary Beulah, who during the project would always pronounce the word robots ro-buts, *we will miss you.*

To Vera Mae, who was the first to ignite Tracey's passion for philosophy and science, you are missed.

About the Authors

Cameron Hughes is a professional software developer with more than 15 years of experience. He is a staff programmer/analyst at Youngstown State University and a software epistemologist for Ctest Laboratories. **Tracey Hughes** is a senior software and graphics programmer at Ctest Laboratories, where she develops information and epistemic visualization software. Both Cameron and Tracey are long-time robot enthusiasts with a collection of more than 100 robots. They have sponsored and participated in local robot competitions and robot programming workshops for the LEGO NXT and RS Media platforms through their local ACM chapter. Cameron and Tracey are the authors of seven books on software development, multithreaded programming, and parallel programming in C++.

Trevor Watkins is a network communications and system integrations specialist. He is currently the Technology Manager at the Wadsworth Public Library, where he designs, integrates, and administers all aspects of the library's network and information systems. Trevor is also an adjunct professor in the Computer Science and Information Systems Department at Youngstown State University, where he teaches high-level programming languages and computer networks. He has been a robot hobbyist for over 20 years, with the past five years dedicated to MINDSTORMS NXT, Vex, and Arduino-based robot kits, and he consults with local high school robotics teams.

Bob Kramer is a full-time computer science professor at Youngstown State University. His research interests include using LEGO robotics as a tool to teach computer science concepts, as well as the development and extension of programming tools for LEGO robots. Bob has helped extend the nxtOSEK environment to enable C++ programs to execute on the NXT platform, and has developed an interface for a third-party sensor in the leJOS environment.

Contents

Introduction

Even though we may be able to get a robot to do many different things and perform different tasks, we will never be able to build a single robot that can perform every task or do everything imaginable. Even a general-purpose robot is limited by the number or types of sensors it has or by the types of end effectors it possesses. We may only have access to a stationary robot where a mobile robot is needed. It might be determined that a four-wheeled tractor has the required type of mobility, and as it turns out, the robot we have is bipedal or has been designed with only two wheels. But we can't go too far in the other direction either. It's not practical or possible to build a different robot for every task or for every scenario we require. First, building a robot requires time, and the parts are costly. Sensors can be expensive. We wouldn't want to build one robot to turn off the lights and a separate robot to turn on the lights. There would be a lot of unnecessary duplication. However, we could dismantle the robot we have to build the robot we need. We really don't like this option, though. After we have put in the time and effort to build a robot and test it, and it does what we want it to do, we're usually happy, and we keep it.

So what is the solution? We don't want to build a new robot for every task that pops up, and we don't want to dismantle a perfectly good working robot that already serves one purpose in order to do some new task. Sometimes we're lucky and we have a third alternative. We might have a robot that seeks and reports the positions of red things, and we might have another robot that retrieves tennis balls. What if we needed a robot that could retrieve apples from the backyard that had fallen from the tree? We could reprogram our tennis ball retrieval robot to look for apples instead of tennis balls. But we still need our tennis ball retrieval robot, so that option is no good. We could reprogram it to retrieve tennis balls and apples, but that would slow it down because now it has to determine whether it is sensing an apple or a tennis ball, and it works just fine at the speed it has. So that's no good. Likewise, we could reprogram our robot that reports red things, but the problem is that robot works just fine the way it is as well.

In this book we present a third alternative: *robot teamwork*. Let's get the robot that finds red things to work together with the robot that retrieves tennis balls to create a robot application that retrieves apples (red ones, of course) from the backyard. This approach does require multiple robots to be involved in the solution, but it has the advantage of not having to totally reprogram or dismantle a working robot.

In this book we introduce the Bluetooth Robotic-Oriented Network (BRON). BRON is a communication technique that allows teamwork between two or more robots. BRON allows robots to share sensors, actuators, end effectors, motor power, and programming in order to accomplish as a team what they could not do individually. Rather than having to totally repurpose a robot or build a completely new robot from scratch, BRON allows you to use existing robots to work in teams to perform tasks that they were not originally designed to perform as individuals. With BRON, we use each robot to do what it is designed to do, and by adding communications between the robots, we can get the team of robots to do new things.

Who This Book Is For

The ideas and concepts presented in this book are for the most part applicable to all the major robot kits available today. Any robot kit that supports Bluetooth and Java, NXC, LabVIEW, or NXTG can be used to build the projects that we present in this book. We chose MINDSTORMS NXT and Tetrix to implement the projects in this book because MINDSTORMS NXT is one of the most widely used and well-known robot kits available. Its relative low cost and the wide availability of add-on sensors make it an easy choice for this book. Because we present the concepts and techniques in this book from beginning design to final detailed implementation, users of other robot kits such as Vex or Arduino can follow along with the projects in this book. But in order to take full advantage of all the examples, readers should be familiar with and have access to a couple of MINDSTORMS NXT/Tetrix robot kits.

The Programming Languages We Use

We present the examples in this book in several languages. National Instruments' LabVIEW and LEGO's NXT-G are used as alternative graphical environments. NXC and Java are used as the nongraphical languages. Although the book does not present every example in each language, the website supporting this book (www.robotteams .org) does have most of the major examples in at least two languages. Downloads for this book are also available at mhprofessional.com/robotteams.

This book is not an introduction to programming in any of the languages. We include several of the most popular languages for programming robots because it is assumed that readers are familiar with at least one of the languages. If you are completely new to programming, then you will need an introductory book that teaches you one of these languages so that you can follow along in this book.

Robot Skill Level Needed

Since most of the robots implemented in this book are simple designs, readers do not have to be advanced robot engineers. But readers are assumed to have some experience in building and programming robots, with a particular emphasis on MINDSTORMS NXT–based robots. This book is not an introduction to robot building. Robot enthusiasts, hobbyists, robotics majors, computer science majors, and those involved in robotics competitions will find the concepts and techniques presented in this book indispensable. Anyone who needs to build a robotic application that requires communication and cooperation between two or more robots will find this book a valuable resource.

The robot teams you will learn how to build and program include

- A crime scene analysis bot
- A robot convoy
- A Rubik's Cube solver

Bluetooth Communications, Programming, and Protocol

You will learn how the Bluetooth communications protocol works and how it is programmed in NXT-G, NXC, LabVIEW, and Java. You will learn how to send and receive Bluetooth messages, data, and commands between robots, between a robot and a computer, and between an Android smart phone and a robot.

Sensors

In addition to learning to build specific robot teams and employ Bluetooth communications, readers will learn how to use light, touch, ultrasonic, compass, color, barometric, and chemical analysis sensors such as a pH sensor.

Unified Modeling Language and Flowcharts

Readers will learn how to use the Unified Modeling Language (UML) and flowcharts to plan communications between robots and how to capture the initial and final designs of robot capabilities.

BRON's Believe It or Not

Most chapters will conclude with a brief section entitled "BRON's Believe It or Not." These sections contain supplementary material that provides interesting, sometimes fun, but little-known facts about robots or robotics. These briefs are not essential in learning how to connect your robots with Bluetooth and can be skipped. In most cases, though, they do provide insight into or a deeper understanding about some area related to robots, robot building, or robot programming.

The Robot Teams Website

This book has a supporting website—www.robotteams.org—that will host complete examples from the book. The examples will be available as NXT-G, Java, LabVIEW, and NXC programs. Because there simply was not room to put the complete build instructions for all the robots we present in this book, and because some of the robots have common builds, we include the complete build instructions along with the parts lists on the website. Also, demonstration videos on each project in action will be available on the website. In addition to complete programming examples, build instructions, and robot demonstration videos, readers will find the latest blogs from the authors and technical articles exploring and discussing the notions, advantages, and disadvantages of building teams of robots. In addition to these, readers will find forums where other robot enthusiasts meet and discuss all things related to robotic teams.

MINDSTORMS and Software Versions

The robot projects presented in this book were built with

- NXT 1.0 and NXT 2.0
- Tetrix for MINDSTORMS

They were programmed in the following environments:

- Mac OSX Lion
- SusE Linux 11.0
- Windows 7

The build instructions were produced using

- Digital Designer 4.2
- ML-CAD 3.30
- LPub 2.4.8.0
- L3P 1.4 Beta 20080930
- POV-Ray for Windows 3.6

The robot teams were programmed using

- LabVIEW 2011 and NXT-G 2.0 for the Macintosh
- Eclipse Helios
- leJOS 0.9.0

The Android smart phone was programmed using

- Ubuntu Linux 12.04
- Eclipse Indigo
- Android SDK, set to version 2.3.3 (API 10)

Robot Building, Testing, and Code Reliability

Although all the robots, examples, and applications in this book were tested to ensure correctness, we make no warranties that the robots, examples, and applications are free of defects or error, are consistent with any particular standard of merchantability, or will meet your requirement for any particular application. These robots and the examples that use them are meant for exposition only. They should not be relied on for solving a problem whose incorrect solution could result in injury to a person or loss of property, time, or ideas. The authors and publisher disclaim all liability for direct or consequential damages resulting from your use of the robots, examples, or applications presented in this book or contained on the supporting website for this book.

Acknowledgments

We could not have successfully pulled this project off without the help, suggestions, constructive criticisms, and resources of many of our friends and colleagues. We would like to thank the students and colleagues who reviewed and gave suggestions regarding the earlier versions of this material; the technical support at Vernier Software and Technology for the equation to convert the pH sensor's raw voltage to the pH measurement; the technical support at LEGO Education, who give us valuable information; and the technical support at Pitsco, who gave us specifications on their DC motors. We are also indebted to the leJOS forums that provided so many nitty gritty ins and outs on Java and MINDSTORMS. Special thanks to Roger Stewart and Patricia Wallenburg, who showed much patience, and Ctest Labs for the use of their Pantheon and robot facilities.

Chapter 1

It Takes Two to Tango

The Lost Scrolls of Robotics: #1
No disassemble!

—Johnny 5, *Short Circuit*

There are *no one-size-fits-all robots*. No matter how hard we try, we cannot come up with a practical robot design that can be used to build a single robot that can perform any and all tasks. No one robot will ever be able to do everything. This is an important fact. There are many reasons this is true. But here are a few:

- Robots have or have access to a limited number of sensors.
- Robots have a limited number of end-effectors and actuators.
- Power requirements limit robot size, weight, strength, speed, and mobility.

The microcontrollers and microprocessors that robots use have designated purposes, communication limits, and calculation limits:

- Sensors have limitations and precision variations and capabilities.
- Different tasks require different types and numbers of actuators.

We may have a task that requires a heat sensor, and our robot is equipped with only a sound sensor. We may have a task that requires that our robot have ultrasonic sensory capability, and our robot is equipped only with color and temperature sensors. The task may require a robot that is capable of navigating vertical surfaces, and our robot has been designed to navigate only horizontal surfaces. Sometimes the requirement is not for a different capability but for a different degree of capability. For instance, we might have a robotic application that calls for three color sensors only to realize that our robot can be equipped with only one color sensor. Or we might have a situation where the robot application needs to be able to push or pull 25 pounds, and we find that our single robot's actuator is capable of moving only 5 pounds.

When the Robot We Have Is Not the Robot We Need

What do we do when the robot we have is not the robot we need? A cloud of depression starts to loom at the mere suggestion of dismantling one of our prize robots. Each robot that we design and build is special, and generally, it has taken a long time to get it exactly where we want it to be, to get the looks cool, the functionality straight, and the programming instructions right. There is usually no interest in any proposition that includes dismantling one of our robots. It's also difficult to find any fans in our group who are interested in reprogramming our robots. Once we've got them programmed, that's it! *If it ain't broke, don't fix it!* However, there is another option. We could just build another robot in this case. In this way, we don't have to worry about dismantling one of our works of art. This works out in some cases—if we have extra parts, microprocessors, and microcontrollers. But building another robot is not always practical. There may not be enough time, enough money, or other necessary resources for the robot build cycle. It simply is not practical to build a new robot every time a new task needs to be performed. So what do we do? In robotics, we have several approaches to this problem:

- Build highly configurable, modular special-purpose robots.
- Build general-purpose robots.
- Build reprogrammable robots.

Special-Purpose Robots Can Be Flexible

Figure 1-1 shows a basic puma robotic arm with several different end effectors. End effectors are devices "at the end" of a robotic arm, designed to allow the robot to interact with its environment. Depending on the task the robot has to perform, the end effector can be changed without having to completely dismantle the robot and without having to build a completely new robot from scratch. The robot in the figure is an example of a configurable robot. End effectors can be expensive and can have very special purposes. The ability to switch off end effectors gives us a much more flexible robot design. Special-purpose robots, although often configurable, are used to perform very specific tasks, and those tasks are typically dedicated to some specific industry, for example, educational, janitorial, bomb disposal, automotive, space exploration, and so on. These kind of robots typically perform one or two types of tasks and offer some configuration so that they can be more flexible in their work. For instance, the robot that is designed to extract dirt from a floor may have one end effector for carpet floors and another end effector for wood floors.

Figure 1-2 shows our basic modular robot arm. That is, not only can it swap out end effectors, but it also can change other major components, such as its base. In Figure 1-1, the robot is a stationary robot. It does not leave its location; although it has moving parts (its robot arm), the robot does not change locations. We can make that

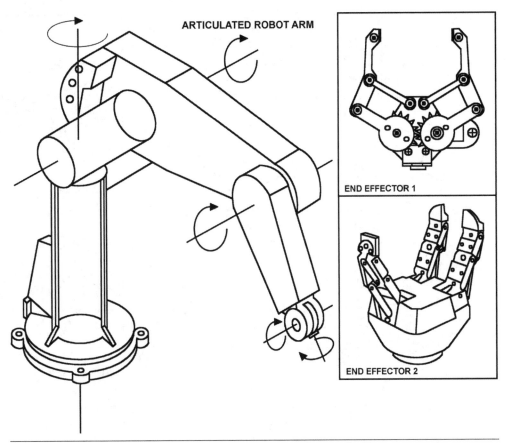

ARTICULATED ROBOT ARM

END EFFECTOR 1

END EFFECTOR 2

FIGURE 1-1   A basic puma robotic arm with two of its end effectors.

robot more flexible by providing a module that has wheels. This will allow the robot arm to solve tasks at different locations. While providing configurable end effectors and plug-in modules increases the flexibility of the robot, is this enough? Will these kinds of configuration options allow our single robot arm to perform any and every task? Although we can change the end effectors so that the robotic arm may grab, turn, drill, remove, push, hammer, and so on, the robotic arm has no visual sensors. It cannot "see" what it needs to act on. We can make the robot arm mobile by giving the robot wheels, but it has to have some way to know where it's going. So if the task is to collect, say, a blue ball and place it into a nearby receptacle, our robotic arm with all its flexibility and mobility can only solve part of the problem. It can lift the ball, it can place the ball in a receptacle, but it has no idea where the blue ball is and no sensors to find it. So in this case the vertical market robot would either have to be rebuilt or extended in some way or we would have to build another robot that knows how to locate things.

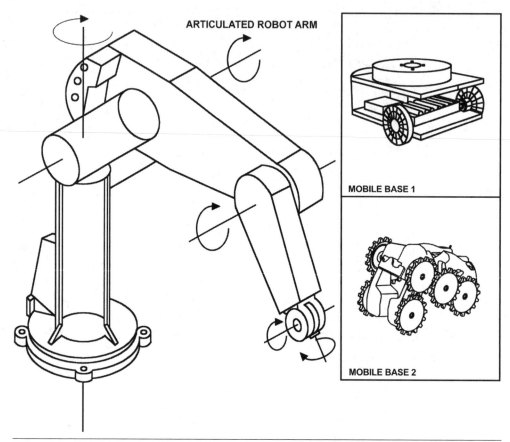

FIGURE 1-2   The puma robotic arm can be configured with different bases that allow the robot arm to be mobile.

General-Purpose Robots: Fact or Fiction?

The general-purpose robot is the dream robot that is the answer to the limitations of special-purpose or dedicated-purpose robots. The basic idea behind the general-purpose robot is that it has not been designed to perform any particular task. So one would think that it has no limitations! Theoretically, a general-purpose robot can do anything that it is programmed to do. The problem is, good luck finding one. So far general-purpose robots are the stuff of movies, sci-fi novels, or video games. Sometimes we confuse robots that are built with no particular task in mind with the notion of general-purpose robots. Figure 1-3 shows a diagram of the Robosapien Class (RS Media) robot and an NAO robot.

This anatomy is often what comes to mind when one thinks of a general-purpose robot. Both the RS Media and NAO robots are directed at the domestic market (i.e., hobbyists, everyday robot lovers, students, robotics enthusiasts, etc.). They are both

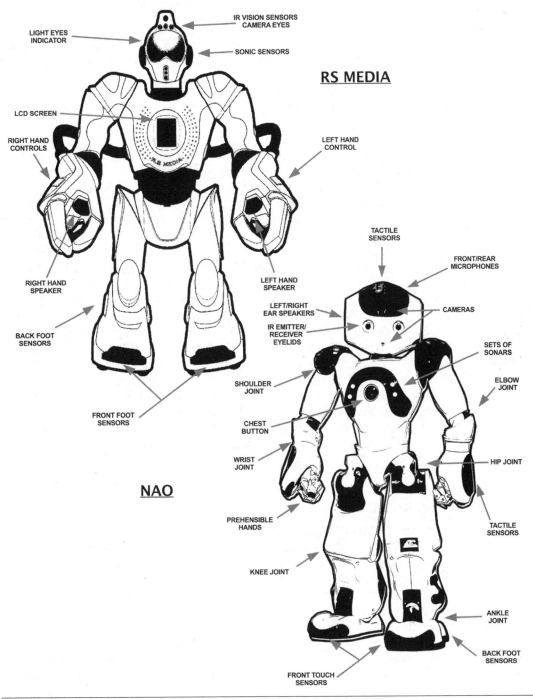

FIGURE 1-3   RS Media and NAO robots showing some of their features, sensors, and actuators.

programmable but haven't been designed to do anything in particular. They both have microcontrollers built in, DRAM, and USB input-output (I/O), and they run the Linux operating system. Pretty generic stuff! Notice in Figure 1-3 that they are both bipedal, have sensors in their heads, and use basic motors for actuators. But what can either of these robots really do? If our goal is a robot that is not hampered by limitations that we can use in all kinds of situations, then robots such as RS Media and NAO should be good news, right? Well, it is good news that they are not programmed just to do one or two specific things. But it is also bad news that they are not programmed to do anything in particular. With a robot that has a robotic arm, we know that the robot's task will be to pick things up, place things, remove things, or insert things. All that's left to determine is, What kinds of things will the robotic arm be working with? How heavy will the objects be? How fragile will the objects be? How high or how low? The answers to these questions determine what types of actuators are needed, how much force the gripper or end effector will need, how long or how short the robot arm needs to be, and so on. If we don't know in advance what tasks the robot will be performing, though, how do we know how much energy output the motors will need, how much force the end effectors will apply, and how big or how small the objects will be that the robot needs to interact with. A robot that has to retrieve eggs or tomatoes requires a different "hand strength," so to speak, from a robot that will be removing bricks from the patio or retrieving cold refreshments. The actuators, power supply, and size and type of end effectors would be completely different for a robot that is handling eggs from a robot that is moving bricks or restraining the family cat. So how do you build a general-purpose robot if you don't know what tasks it will be asked to perform? What would a truly general-purpose robot look like? If it is truly general purpose, then it should be able to navigate stairs and traverse vertical and horizontal surfaces, and it should have the capacity to do air reconnaissance and underwater roving. What would such a robot look like? Would it be bidpedal like RS Media and NAO? What would it have for end effectors? Hooks, fingers, grippers, hands? Would it have pneumatic or servo-driven actuators?

Reprogrammable Robots

We are very skeptical about the notion of general-purpose robots. We haven't really seen any that deliver on the promise. However, we have seen multipurpose robots—robots that can do many things as opposed to specialized robots that can do only a few things. With multipurpose robots, the tasks may vary widely; the multipurpose robot is flexible but is limited by its sensors, actuators, microcontroller, and end effectors. One of the distinctions between a general-purpose robot and a multipurpose robot is that you can tell right away what a multipurpose robot cannot do. It cannot sense beyond its sensors. It cannot lift objects beyond the capacities of its actuators. It cannot process more or faster signals than its microcontroller. There is no suggestion that a multipurpose robot can do everything, whereas general-purpose robots promote the illusion that they can do any task.

Flexible Special-Purpose Robots and Reprogrammable Multipurpose Robots

The configurable special-purpose robot can and often does get the job done. But it has obvious limitations. The reprogrammable multipurpose robot can get even more done, but it, like all robots, will forever be limited by its access to sensors, actuators, and end effectors. So although a multipurpose robot can be used to perform many tasks, it cannot do them all. And to emphasize one of the major points of this book, it is not practical to build a new robot for every new task that comes up.

So what do we do when the robots we have are not the robots we need? What do we do when our robots are constrained by some limitation? What do we do when our robot almost solves the problem, or almost has the strength that it needs, or almost has the sensory capability that the task requires?

Two Microcontrollers Are Sometimes Better Than One

In this book we show you how to take advantage of the age-old concept of teamwork. You know, the old "two heads are better than one" concept. We show that two robots together can sometimes accomplish what neither one of them could accomplish alone. We demonstrate how a little extra planning before building your robot will allow your robot to work as a standalone robot performing the task you designed it to perform or as a team member helping to perform a task that you would never have guessed your robot would be involved in. In this book we show you how to extend a robot's sensors' limitations, power limitations, or computational limitations through teamwork. The teamwork is made possible by a communication method named *Bluetooth* (more on this later). This book is about taking robots that can perform simple single tasks and putting them in teams of two or three and then getting them to perform more complex robotic applications based on the single tasks they are already programmed to perform. In other words, we show you how to repurpose your robots without major rebuilding or reprogramming.

We also show you the concept of factoring your robot designs into teams from the very start of the design process. Instead of designing a single complex robot with lots of sensors that perform lots of tasks, we suggest designing the application so that it will consist of two or three simpler robots working together as a team or unit. Sometimes it's easier to program simple robots to do simple tasks and later team up to perform more complex tasks than it to try to make one big "jack of all trades" robot that is complicated to program, complicated to test, and complicated to get going and keep going. In this book we will show you how to build robot teams (*collectives*) based on Bluetooth Robotic-Oriented Network (BRON) techniques. Because the Bluetooth technology is used by other devices, such as computers and smart phones, we will show you how to build robot-oriented teams that contain a mix of devices that work together with your robots. The only thing better than designing and building a robot to do a job is building a robotic team, a robotic collective, that will be far more capable than any single robot. However, in cases where a single complex robot consisting of multiple microcontrollers is the better choice, we will show you how to use Bluetooth communication techniques to coordinate the functions of such a complex robot.

Possible Teams, Possible Players

In a BRON, the robot is the central player, but the network can contain other robots or devices. The important point here is that the robot and the other networked devices all work together as a team to make a single robotic application. We will introduce several network or team configurations. In some cases, the robot will explicitly control the other devices, and in other cases, the other devices will control the robot. For example, in some robot team configurations, we use Bluetooth-enabled computers to send commands to control NXT-based robots, and in other configurations, commands are sent to and from devices such as the Android-based smart phone. For instance, with the robot and Android phone configuration, the robot may direct the smart phone to send a text message. Table 1-1 lists the potential team members that we will use for our BRON. The table also shows the microcontroller or microcomputer capabilities of all the team members involved.

The *team capability matrix* (TCM) is something that is constructed before building any robots or trying to design a robot application. The TCM is used to identify what Bluetooth-enabled devices are potential team members and what set of capabilities each device brings to any solution or robot design. It's a good practice to construct a TCM prior to getting started on your team of robots. Once a TCM is constructed, it should be maintained. An up-to-date TCM will make it easier to build new projects and also will keep the full capability of your current project easily accessible. Notice in Table 1-1 that NXT robots bring to the table such sensors as a compass, color, and so on. While smart phones bring high-definition cameras, Internet connections, and Global Positioning Satellite (GPS) capabilities, computers bring multicore processors, high-speed computation, massive storage devices, and wireless connectivity. Each potential team member has something to offer that typically the other team members do not have. This is another benefit of using Bluetooth to build networked robot applications. The robots can access capabilities that are not part of the initial robot kit. The capabilities of any particular device may drive the configuration of the network. Because the NXT and Tetrix robot is capable of a wide variety of end effectors and the widest variety of sensors, it might be at the center of the configuration, or if intense symbolic computation is a major part of the robot application, the computer may be at the center of the team. Figure 1-4 shows the major teamwork configurations or teamwork topologies that we work with in this book. While these are not the only possible topologies, they are some of the most important ones in use.

There are two basic types of robot teams:

- Homogeneous teams
- Heterogeneous teams

TABLE 1-1 Team Capability Matrix (TCM) Showing the Capabilities of the Potential Team Members We Will Use for the BRON

Team Members (Devices)	Microcontroller / Processor	End Effectors	Mobility Type	Display	Sensors	Motors/ Controllers	Communication	Other
Team D								
Robots	4 × 32-bit ARM7 NXT bricks	1 × arm and gripper 1 × arm with probe effector	1 × 4 wheel 2 × tractor wheels 1 × 3 wheels	4 × 100- × 64-pixel LCD screen	2 × LEGO touch 1 × HiTechnic 16 color 1 × LEGO color 2 × LEGO ultrasonic 1 × LEGO sound 1 × HiTechnic compass 1 × LEGO temperature 1 × Vernier pH sensor 1 × CODATEX RFID system	4 × Tetrix dc motors 3 × Tetrix servos 2 × Tetrix dc motor contollers 2 × Tetrix servo motor contollers 6 × LEGO servo motors	4 × Bluetooth 4 × USB	2 × Tetrix battery packs 2 × Tetrix power switches
Devices	AMD Turion 64-bit dual 2.8-GHz processor (laptop)	N/A	Portable	19-in HD LCD	N/A	N/A	802/11b wireless Bluetooth USB 10/100 Ethernet modem infrared	1.5 TB of storage

(continued on next page)

TABLE 1-1 Team Capability Matrix (TCM) Showing the Capabilities of the Potential Team Members We Will Use for the BRON *(continued)*

Team Members (Devices)	Microcontroller / Processor	End Effectors	Mobility Type	Display	Sensors	Motors/ Controllers	Communication	Other
	32-bit dual-core ARM (Android)	N/A	N/A	800 × 400 WVGA	Camera	N/A	N/A	GPS, gyroscope, etc.
Team C								
Robots	3 × 32-bit ARM7 NXT bricks	1 × arm and gripper	3 × tractor wheels	3 × 100- × 64-pixel LCD	3 × LEGO ultrasonic 2 × LEGO color 1 × LEGO touch	7 × LEGO servos	3 × Bluetooth 3 × USB	
Team B								
Robots	3 × 32-bit ARM7 NXT bricks	1 × flipper 1 × gripper 2 × rotating platforms 1 × aligner with fingers	N/A	3 × 100- × 64-pixel LCD	N/A	5 × LEGO servos	3 × Bluetooth 3 × USB	
Devices	32-bit dual-core ARM (Android)	N/A	N/A	800 × 400 WVGA	Camera	N/A	N/A	GPS, gyroscope, etc.

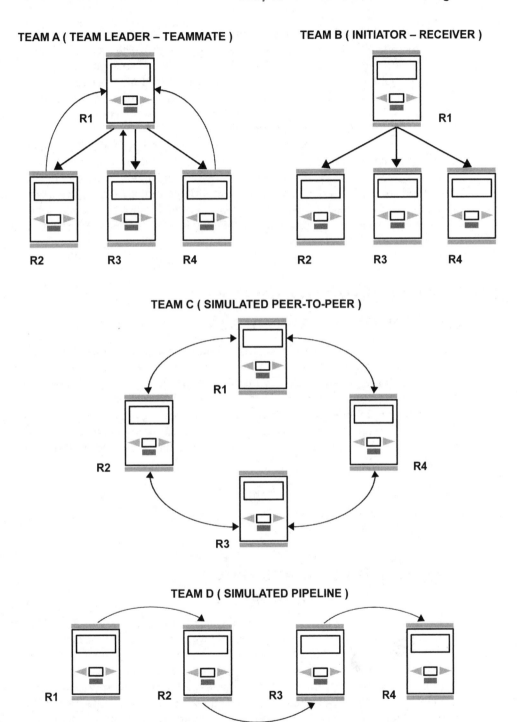

FIGURE 1-4   The major teamwork topologies used in this book.

Homogeneous robot teams consist of two or more robots that have the same type of microcontroller. The robots may be a mix of tractor-based, stationary, bipedal, etc. robots as long as they have the same microcontroller; for our purposes, they are in the same class. If the BRON consists of only MINDSTORMS NXT robots, then we have a homogeneous robot team. *Heterogeneous robot teams* consist of two or more robots with either different types of microcontrollers or one robot connected with one or more nonrobotic Bluetooth-enabled devices.

While MINDSTORMS NXT–based homogeneous robot teams are capable of accomplishing incredible feats, a heterogeneous robot team is far more flexible and capable of accomplishing even more fantastic feats. We will show you how to build both kinds of teams in this book. The Bluetooth technology (which we discuss in detail in Chapter 2) is the magic behind both types of robot teams. Bluetooth allows the robots to send and receive commands and data from other NXT robots or other compatible Bluetooth devices.

Do Networked Robots Equal Robot Teamwork?

Two or three connected NXT microcontrollers do make a network. But does a network automatically mean that we have a team? Unfortunately not. If only making the Bluetooth connection did produce a team, life would be so much easier. In our case, the Bluetooth connection is what makes a network of robots possible, and the network is what makes a team possible. You can have a network with no teamwork. Teamwork is the result of organizing and coordinating the team members to accomplish some task together. There are three basic ways to coordinate two or more robots or team members:

- Time- or chronology-based coordination
- Event-based coordination
- Message-based coordination

Coordinating Robots Based on Time or Chronology

Let's say that we have three robots. We'll call them Robot A, Robot B, and Robot C. Each robot has some unique capability, and we wish to use these robots together as a team. In time- or chronology-based coordination, Robot A, Robot B, and Robot C are synchronized based on time or chronology. For example, Robot A starts a task at precisely 1200 hours. Robot B will not start its task until precisely 15 seconds from whatever time Robot A started. Robot C will start its task only after Robot B has been operating for 10 seconds. Or the coordination might be completely dependent on chronology. For example, Robot B and Robot C both could start at the same time; that is, regardless of when they start, they both start together. Robot A is allowed to start only after either Robot B or Robot C has finished. In chronology-based coordination, the order in which the robots start and finish is critical. There are nine possible order configurations between any two robots or team members. Table 1-2 lists the possible configurations.

TABLE 1-2   List of Possible Chronology-Based Configurations Between Robots

Robot A	Robot B
Start	Start
Start	Finish
Finish	Start
Finish	Finish
Start	Wait
Wait	Start
Finish	Wait
Wait	Finish
Wait	Wait

For example, the table says that Robot A and Robot B can have a *start-to-start* (SS) coordination. This means that Robot A and Robot B have to start at the same time. A *start-to-finish* (SF) coordination means that Robot A cannot start until Robot B finishes. A *start-wait* (SW) coordination means that Robot A cannot start until Robot B waits or Robot A can start only if Robot B waits. Thus, if we say that two robots have a *finish-to-finish* (FF) relationship, this means that no matter what happens, both robots have to finish at the same time. These coordination relationships are important for a number of reasons. But this is usually easy to see if you ask yourself, When this robot is performing this task, what are the other robots doing? Should they be doing anything? Does Robot A depend on Robot B for some reason? While time or chronology coordination may appear straightforward on the surface, we shall see that there are many pitfalls. The robots to be coordinated are operating with different motor speeds and different acceleration settings. Power source can become a factor. Things such as wheel size can affect distance-versus-time calculations. Even size and weight can become a factor when attempting to coordinate two or more robots based on time. To see what it looks like to coordinate two robots by time, let's take a look at Listing 1-1 and Listing 1-2.

LISTING 1-1   Coordinating two robots by time using leJOS.

```
1    // Listing 1.1  example of time coordination for A1R1
2    import lejos.nxt.*;
3
4    public class test_motor{
5
6    public static void main(String[] args) throws Exception
7    {
8
9
10       Thread.sleep(5000);
11       LCD.drawInt(Motor.A.getAcceleration(),0,0);
12       LCD.drawInt(Motor.B.getAcceleration(),0,1);
13       LCD.drawInt(Motor.A.getSpeed(),0,2);
14       LCD.drawInt(Motor.B.getSpeed(),0,3);
15
16
17       Motor.A.setSpeed(360);
18       Motor.C.setSpeed(360);
19       Motor.A.forward();
20       Motor.C.forward();
21       Thread.sleep(3000);
22       Motor.A.stop(true);
23       Motor.C.stop(true);
24       Thread.sleep(5000);
25       Motor.A.forward();
26       Motor.C.forward();
27       Thread.sleep(2000);
28       Motor.A.stop(true);
29       Motor.C.stop(true);
30       Thread.sleep(5000);
31       Motor.A.forward();
32       Motor.C.forward();
33       Thread.sleep(6000);
34       System.exit(0);
35
36    }
37
38    }
```

LISTING 1-2   Coordinating two robots by time using NXC.

```
/* Listing 1.2  example of timed coordination for A3R2 */

 1  task main(){
 2      Wait(5000);
 3      NumOut(0,LCD_LINE1,MotorActualSpeed(OUT_B));
 4      NumOut(0,LCD_LINE2,MotorActualSpeed(OUT_C));
 5      OnFwdSync(OUT_BC,50,1);
 6      Wait(2000);
 7      NumOut(0,LCD_LINE3,MotorActualSpeed(OUT_B));
 8      NumOut(0,LCD_LINE4,MotorActualSpeed(OUT_C));
 9      Off(OUT_BC);
10      Wait(5000);
11      OnFwdSync(OUT_BC,75,-1);
12      Wait(3000);
13      NumOut(0,LCD_LINE5,MotorActualSpeed(OUT_B));
14      NumOut(0,LCD_LINE6,MotorActualSpeed(OUT_C));
15      Off(OUT_BC);
16      Wait(5000);
17
18  }
```

Listing 1-1 is a simple Java program that controls a robot that we name A1R1, and Listing 1-2 is a simple C program that controls a robot that we name A3R2. All of the robots in this book have names that are four characters long. Fortunately, there is a method to our naming madness. Table 1-3 explains the robot naming conventions that we use.

TABLE 1-3   The Robot Naming Conventions We Use in This Book

Character Position	Explanation
1	Team name
2	Stage of development (robots may have up to five stages in this book)
3	Robot or device (R or D)
4	Team member number

So A3R2 is a third-stage robot on A team and is the second member of the team. Let's take a closer look at what A3R2 is programmed to do.

A1R1 and A3R2 have an SS relationship in this example. That is, they both start at the same time. Notice line 2 in Listing 1-2. We have

```
Wait(5000);
```

This means that A3R2 is going to wait 5 seconds before doing anything; 1000 = 1 second; 2000 = 2 seconds; and so on. A3R2 waits 5 seconds and then sends its actual motor speed to the display. A1R1 also waits 5 seconds with

```
Thread.sleep(5000);
```

This brings us to an important benefit of networked robots and robot teams. The members of the team can be programmed in different languages and still may work together. In fact, the programming language can be the benefit or capability for which a particular robot has been added to the team. It may be the case that we have two or three robots already built. One may be programmed using the NXT-G, and another may be programmed in Java or C. We might wish to take advantage of some nice Java capability, for example, built-in navigation libraries, and we may not want to rebuild or flash the firmware of any of our other robots, so we choose the robot that is already running leJOS (Java-based firmware) because of the navigation libraries that leJOS NXT has. Team members still can work together, even though they've been programmed in different languages. In Listing 1-2, motors B and C are being synchronized for A3R2 on line 5. The command

```
OnFwdSync(OUT_BC,50,1);
```

causes motors B and C to be synchronized and move forward. It sets the motor power to 50 and the turn percent of A3R2 to 1. The speeds of the NXT motors are measured in a range from 0 to 100. A setting of 0 means that the motor is not spinning at all, and a setting of 100 means the motor is operating at 100 percent, or as fast as it can. We can think of the power parameter as representing the percent of speed at which the motor is operating. So on line 5 of Listing 1-2, A3R2 is being set to run at 50 percent and with a 1 percent turn. The turn percent is measured from –100 to 100.

Figure 1-5 shows how the turn percent is divided up. A value of 0 means that the robot is moving in a straight line or relatively straight line. Positive values greater than 0 up to 100 determine how sharp the movement to the right is. Negative values from –1 to –100 determine how hard to the left the robot is moving.

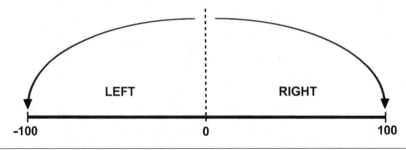

FIGURE 1-5 The turn percent diagram.

The next command

```
Wait(2000);
```

really tells the microcontroller not to send any more instructions to the robot for 2 seconds. This is kind of confusing at first. It looks like we are commanding A3R2 to wait for 2 seconds. This actually has the opposite effect. Because A3R2 will not receive any further instructions for the next 2 seconds, this means that it is going to keep doing whatever it was doing prior to the `Wait()` command. In this case, it was moving forward with 50 percent motor power with a turn percent of 1 to the right. Recall in line 2 of Listing 1.2 that we had

```
Wait(5000);
```

This makes it look like we are commanding A3R2 to wait for 5 seconds. But remember that the `Wait()` command allows A3R2 to continue doing whatever it was doing prior to sending the `Wait()` command. In this case, since A3R2 was doing nothing prior to the `Wait()` command, it will continue to do nothing for 5 seconds. So we achieve the goal that we wanted. Just remember that `Wait()` will cause the flow of control to pause for whatever time interval you give it (measured in seconds), so whatever the robot was doing right before the call to `Wait()`, it will continue to do for that time interval.

On lines 7 and 8 of Listing 1-2, we have

```
NumOut(0,LCD_LINE3,MotorActualSpeed(OUT_B));
NumOut(0,LCD_LINE4,MotorActualSpeed(OUT_C));
```

These commands cause the current speed of A3R2 to be written to A3R2's liquid-crystal display (LCD). In our case and on the first run of this program, the LCD had the values of 50 and 45, and the second run of the same program had the values 60 and 44. This means that the power was at 60 for motor B and 44 for motor C. We point out the motor speeds here because depending on the surface, the power supply, and the calibration of the motors for the NXT, you may get different values.

```
OnFwdSync(OUT_BC,75,-1);
```

On line 11 of Listing 1-2, we set A3R2's motor to 75 percent, with A3R2 moving forward and slightly to the left. After sending the actual speed to A3R2's LCD, we get 76 and 67 for the first run and 76 and 70 for the second run. For some reason, motor B is running at or above the 75 speed, and motor C is running slower. We will go into the reasons and solutions for these kinds of deviations throughout this book. Recognizing and accounting for these kinds of issues are important parts of coordinating your robot teams.

A3R2's and A1R1's Time Coordination

The `Thread.sleep()` command (from Java) and the `Wait()` command (from NXC-C) are both used to coordinate the two robots based on time. They both start processing together. The goal is to have them end at the same place, even though they have different motor speeds and will be traveling for different amounts of time. A1R1 is directed to travel at 360 degrees per seconds with the commands

```
Motor.A.setSpeed(360);
Motor.C.setSpeed(360);
```

for 3 seconds, then for 2 seconds, and finally for 6 seconds. A3R2 is directed to travel at 50 percent and then 75 percent of motor speed for 2 seconds and then 3 seconds. We would like A3R2 and A1R1 to start at the same time and finish in the same place. TimeCoordination.mov shows how successful these programs are. Time coordination is important but can be difficult, especially if the precision of the motors and actuators does not meet the necessary threshold. To strengthen coordination, let's look at the second form of robot team coordination, event-based coordination.

Event-Based Robot Coordination

Robots also can be coordinated based on events. The relationships shown between two or more robots in Table 1-2 also hold for event-based coordination. Robot A and Robot B may wait for some event to occur to take action. Robot A may start command execution on some event, and Robot B might end command execution on the same event. For example, Robot A may be a search robot that is programmed to find an object and report the location of that object. Robot B may be a retrieval robot. Robot B is given the location of the object and then retrieves it. Robot B cannot do its job until Robot A does its job. The event that causes Robot A to stop processing is the location of the object, and the event that causes Robot B to start processing is the location of the object.

But What Is an Event?

Every robot operates within a particular physical environment. The environment will have a finite three-dimensional (3D) space. Okay, to be more precise, the robots and the environments that we deal with in this book are limited to a finite 3D space.* Any action, occurrence, or change that takes place within the robot's environment can constitute an event if the robot has some way to perceive, sense, or interact with it. Now, if something takes place in the robot's environment and the robot has no way to perceive, observe, measure, or interact with it, then it is not the type of event that will concern us. For programming purposes, we typically classify the events as either

---

*We cannot confirm or deny the existence of robots that are not limited to three-dimensional space.

internal events or external events. *Internal events* are events that happen within the robot, for example, changes in motor speed or changes in signal strength of a sensor. *External events* are changes in the robot's external environment. The temperature can rise or fall. An object may change location or color. A sound may start or a continuing sound may stop. The ambient light in the robot's environment may change, and so forth.

In the context of programming a robot team or a robot network, the internal and external events are further divided into local and global events. *Local events* are events that are only available to individual members of a team. For example, if the motor speed of Robot A changes, Robots B and C are not aware of it unless Robot A decides to communicate in some way that the motor speed has changed. *Global events*, on the other hand, are events that are available in some measurable way to all the members on the team. Although the team members may perceive, sense, or interact with the event differently, it is available to everyone. Figure 1-6 shows how the event classification for robot teams can be divided into four quadrants.

As we shall show throughout this book, it is important to understand the classification of events as they relate to building a team of robots that work together to execute tasks and solve problems. The classification of events has three primary ramifications:

- Not all robots have the same sensor or actuator suite. Some events will be perceivable by some robot(s) on the team, whereas those same events will go totally unnoticed by other robots on the team.
- How an event is communicated among the team of robots will depend on whether the event is local or global or internal or external.

	LOCAL	GLOBAL
INTERNAL	**I** Events that cause internal change; e.g., state changes in memory, sensors, motors that cannot be perceived by other robot teammates	**II** Events that are internally shared by all robots; e.g., sense of time, gravitational pull, etc.
EXTERNAL	**III** Events that cause changes in the robot's environment or that are external to the robot that cannot be sensed or observed by other robot teammates	**IV** Events that are external to all robot teammates; e.g., changes in external environment of the robot that can be sensed, observed, or otherwise available to all robot teammates

FIGURE 1-6 Event classification for robot teams.

- The fault tolerance of the robot application will depend on the distribution of events, how they are communicated, and which robots process which events.

For example, quadrant I represents the events that are internal and local to a particular robot on the team. These events are often the cause of entire team failure. Take, for instance, some event fails to happen, and that event is only supposed to happen within the environment of Robot A and is only perceivable by Robot A, and let's say that Robot A doesn't have any special programming to deal with this. Then the whole team that is in some way depending on Robot A's event is left out in the cold, and no one knows what to do. The whole application comes to a screeching halt because no one except Robot A can even perceive the event, and Robot A has failed to do so. If there is no way to recover from this situation, then the robot application has very little fault tolerance. Preparing for problems with the event in quadrants I and III is especially critical to the fault tolerance of the entire team. Problems with the events in quadrants I and IV are also important, but because there is at least a moderate amount of redundancy with the events in quadrants II and IV, there are more ways to fix the problem. We shall return to the issue of robot application fault tolerance throughout this book mostly because the occurrence and communication of events constitute one of the most important, if not *the* most important, method of coordinating a team of robots. The fault tolerance of an application depends on how robust event handling is within the team. And since not all events are created equal, we will need different programming techniques and error-recovery techniques to handle the various event types.

A Simple Event Coordination Between Two NXT Robots

To get started, let's use our two NXT robots, A2R1 and A3R3, to show a simple team coordination based on a local but external event. The flowcharts in Figure 1-7 show the processing for both robots. A2R1 will sense the local external event and communicate it to A3R3.

Notice that both robots in the flowcharts have a team mode. We will cover this in detail later. Team mode indicates that the robots can operate separately or as part of a team. Recall that part of the value of the team approach is that we are able to take robots that were originally programmed to go solo and have them participate in tasks and solve problems that they were not originally programmed to do. A2R1 is programmed only to check for red objects. That's all it does in life. Each day at the same time it moves forward to location A to see if a red object is there. If it is, the robot will add one to its counter of red objects; otherwise, it goes home. It will check only five times. If there is no red object by the fifth check, it will go home for the day and try again tomorrow. Robot A3R3 is programmed to retrieve objects from location A. It doesn't know what kind of object, the color of the object, or even when the object will be at location A. It only knows to go to location A to get the object when told to do so. In our simple example, we would like to retrieve only red objects. Well, A2R1 knows how to identify red objects. A3R3 knows how to get objects from location A. If there were some kind of way to identify red objects at location A, we could combine these two robots in team mode to solve our problem. And that is precisely what each robot is directed to do in

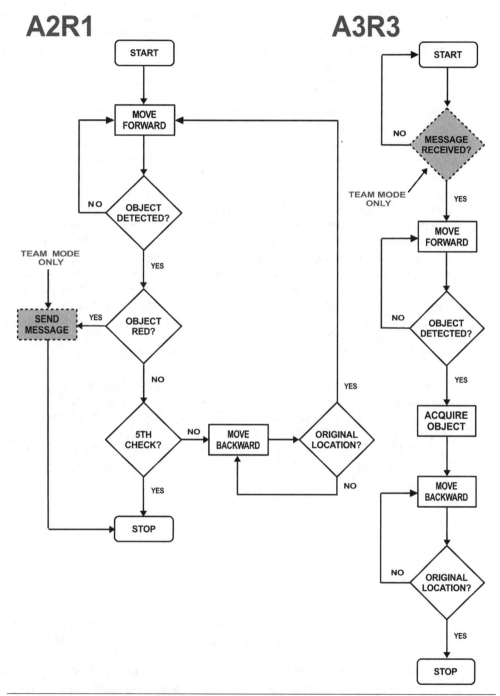

FIGURE 1-7   Simple example of team coordination based on an event. When A2R1 encounters a red object, A3R3 acquires it.

Figure 1-7. The figure has flowcharts that capture the teamwork of A2R1 and A3R3. Throughout this book we will use flowcharts to pictorially describe the tasks that a robot component, a robot, or a robot team is supposed to execute. If you are unfamiliar with the meaning or need to refresh what you learned about flowcharts, the symbols and their meanings are listed on our website at www.robotteams.org/symbols.html.

Why Use Flowcharting?   We use flowcharting as a design and communication tool. It's usually easier to picture the logic of what you want your robot to do first and work out the tasks pictorially. This means having a good understanding of what you want your robot or robot team to do first and then code the instructions in Java, NXT-G, LabVIEW, or C. If you're working with others, flowcharting is a *language-neutral* way to describe the tasks that you want the robots to perform. Keep in mind that not everyone programs their NXT creations using the same language. And as the robot programming in Listings 1-1 and 1-2 demonstrates, robots on the same team may be programmed with different languages. Flowcharting is a way to show what you want the robot to do without getting bogged down in the particulars of a programming language. In addition, it saves batteries for the NXT brick. Rather than writing code, loading it, watching the robot do the wrong thing, then making a change or two to the code, uploading it, and watching the robot do the wrong thing again, flowcharting lets you visually inspect what you think you want the robot to do. In many situations you will be able to catch the problem in the flowchart before ever writing a line of code. For example, what do A2R1 and A3R3 do if they are not in team mode?

The Local External Event   The local external event in this case is the object being red when A2R1 is inspecting it. In our example, we place a green ball at location A. We can arbitrarily replace it at any time with a red ball. If the ball is green when A2R1 inspects it, then it is ignored, and A2R1 goes home. However, if the ball is red, then A2R1 sends a Bluetooth message to A3R3, and A3R3 retrieves the ball. This is a simple example of teamwork. What are the capabilities that are involved?

- A Hi-Technic color sensor
- Five motors (actuators)
- A sound actuator
- A touch sensor
- Bluetooth communications
- Java leJOS
- NXC

But to make it clear which team member can contribute what, we use our team capability matrix (TCM) during the design of our simple solution. The TCM is shown earlier in Table 1-1.

The local event (object is red) cannot be perceived by A3R3. A3R3 does not have a color sensor or any way of measuring "red." So, when the local event occurs, it is time for the teamwork to kick in, and A2R1 must signal to A3R3 that it's time for A3R3 to do its part. Listing 1-3 shows the Java code.

LISTING 1-3   Java code executed by A2R1 that looks for a red object and, if one is found, then notifies A3R3.

```
 1   //Listing 1-3  Java code executed by A2R1 that
 2   //looks for a red object and if one is found then
 3   //A3R3 is notified using Bluetooth communications
 4
 5
 6   public boolean isRed() throws Exception
 7   {
 8
 9      System.out.println("checking is red");
10      if(CurrentColor.getColor()  == CurrentColor.RED)
11      {
12
13        if(TeamMode == true){
14            A3R3 = Bluetooth.getKnownDevice("A3R3");
15            A3R3Connection = Bluetooth.connect(A3R3);
16            if(A3R3Connection == null){
17                LCD.drawString("Where's  A3R3",0,0);
18                Sound.beepSequence();
19                Thread.sleep(1000);
20                Sound.beepSequence();
21                Thread.sleep(1000);
22                System.exit(5);
23            }
24            else{
25                  LCD.clear();
26                  LCD.drawString("sending red",0,0);
27                  DataOutputStream Dout;
28                  Dout = A3R3Connection.openDataOutputStream();
29                  Dout.writeChars("red");
30                  Dout.flush();
31                  LCD.clear();
32                  LCD.drawString("Red Sent",0,0);
33
34                  Dout.close();
35                  moveAwayFromObject();
36                  A3R3Connection.close();
37
38            }
39        }
40        Sound.beepSequenceUp();
41        return(true);
42    }
```

```
43          Sound.beepSequence();
44          return(false);
45
46
47  }
```

Making a Simple Bluetooth Connection   The Java method isRed() is part of A2R1's program that detects red objects. If A2R1 detects a red object, it connects to A3R3 using the following Java methods:

```
A3R3 = Bluetooth.getKnownDevice("A3R3");
A3R3Connection = Bluetooth.connect(A3R3);
```

from lines 14 and 15 of Listing 1-3. These methods are care calls to the Bluetooth class that comes with the leJOS Java environment. We will have a closer look at leJOS in Chapter 6. Once this connection is made, A2R1 and A3R3 form a BRON. They can then communicate with each other. Notice in lines 14 and 15 that the robot's name can be used to connect. Each NXT microcontroller can be given its own unique name. The default name is usually NXT. We have named the robots in this example: A2R1 and A3R3.[*] In this case, A2R1 makes A3R3 aware that a local event has occurred and sends A3R3 a message using

```
Dout = A3R3Connection.openDataOutputStream();
Dout.writeChars("red");
Dout.flush();
```

from lines 28, 29, and 30 of Listing 1-3.

These methods open a connection and send the message "red" to A3R3. When A3R3 is operating solo, its normal operation is simply to retrieve an object from location A. However, when A3R3 is in team mode, it will wait until it receives a message, and then it will retrieve some object from a location A. Listing 1-4 shows the Java code.

LISTING 1-4   Java code executed by A3R3 that waits for a connection from A2R1.

```
1  // Listing 1-4  Java code executed by A3R3 that
2  // waits for a connection from A2R1 if TeamMode is true
3  // A3R3 will only call getXMessage if in TeamMode
4  // and has received a message from A3R3
5
6    public retrieved_object(boolean Tm)
```

---

[*]See robot naming conventions earlier in this chapter (Table 1-3).

```
 7  {
 8      TeamMode = Tm;
 9      if(Tm == true){
10          LCD.drawString("Connecting....",0,0);
11          A2R1Connection = Bluetooth.waitForConnection();
12          LCD.clear();
13          LCD.drawString("A2R1 Found",0,0);
14      }
15      TSensor  = new TouchSensor(SensorPort.S1);
16      SensorPort.S1.addSensorPortListener(this);
17
18
19
20  }
21
22
23  public void getXMessage() throws Exception
24  {
25
26      A2R1 = A2R1Connection.openDataInputStream();
27
28      XMessage = new String(A2R1.readLine());
29      LCD.clear();
30      LCD.drawString(XMessage,0,0);
31      A2R1.close();
32      Thread.sleep(200);
33      A2R1Connection.close();
34
35  }
```

Waiting for a Connection Before Proceeding   A3R3 waits for a Bluetooth connection using

```
A2R1Connection = Bluetooth.waitForConnection();
```

from line 11 of Listing 1-4. This method causes A3R3 to block or to wait until it gets a connection from A2R1. Notice that A3R3 will only wait for a connection if it is in **TeamMode**. If A3R3 is in **TeamMode** and it receives a connection, then it will use the **getXMessage()** method from lines 23 through 35 of Listing 1-4 to get the message from A2R1. The message is read from the Bluetooth connection using

```
A2R1.readLine();
```

from line 28 of Listing 1-4. This causes A3R3 to read a string message.

Teamwork and the Local External Event   We cover the Bluetooth read and write methods in more detail in Chapter 2. Although A3R3 does not sense the local external event experienced by A2R1, the communication channel between the robots allows them to coordinate based on the local external event. Coordination of teamwork based on local external events is at the heart of BRON teamwork because in most cases each team member brings unique sensors and actuators to the table. Not every robot will be able to sense or perform the same actions, so the ability of one robot to communicate with another robot that some local external event has taken place allows both robots to take some kind of action, even though only one of the robots actually experienced the event. The communication (in this case Bluetooth communication) allows all the robots on the team to essentially share sensors and actuators. While this kind of communication and sharing supports powerful multiagent and distributed programming schemes, it comes at the cost of being more complicated than programming robots that go solo. For example, the constructor `retrieved_object()` on lines 6 through 17 of Listing 1-4 shows that A3R3 waits for a Bluetooth connection if it is in **TeamMode**. But what if A3R3 is in **TeamMode** and waiting on a Bluetooth connection from A2R1, and A2R1 fails before making the connection and sending the message for which A3R3 is waiting. Will A3R3 wait on the connection until it exhausts its power source? Or what if A3R3 and A2R1 actually make a connection, but A2R1 fails to send any message or fails to send the correct message? What action should A3R3 take? Will it take any action? Is there any appropriate action for A3R3 to take if A2R1 fails to connect or fails to send the correct or complete information? Answering these kinds of questions and devising the appropriate course of action for all team members when one or more team members are in a failed state constitute the central and most challenging efforts involved in programming teams of robots to perform tasks collectively. We introduce techniques and approaches throughout this book aimed at tackling these challenges.

Message-Based Coordination

In addition to time- or chronology-based coordination and event-based coordination, message-based coordination is the third basic method used to accomplish teamwork among a collection of robots. Although we introduce these three basic method separately, it is important to note that they are often used in combination. Event-based coordination may be used in conjunction with message-based coordination, and message-based coordination may be used in conjunction with time-based coordination, and so forth.

So What Is a Message?

A *message* can be anything from a single character or number to a string containing any kind of information. In fact, any piece of data or information exchanged between two or more robots can be considered to be a message. We use several important message-passing schemes or paradigms to accomplish our robotic teamwork:

- Initiator → receiver (sender-receiver)
- Producer → consumer
- Team leader → team member
- Peer ↔ peer
- Client ↔ server

The message-passing scheme is closely related to the control scheme between team members. Some team members are primarily responsible for generating and sending messages, whereas at times some team members only receive and act on messages without sending messages. These are often called *producer-consumer* or *initiator-receiver relationships*. It's often (though not always) the case that the producer or the initiator is in control of the application and determines when it begins and when it ends. In the cases where consumers or receivers do send information, it is usually restricted to acknowledgment, verification, response, or summary-type information. As we shall see when using NXT microcontrollers, the initiator of the Bluetooth connection is often the producer or sender of the primary messages. In the peer-to-peer configuration, control of critical information can go both ways. The information exchanged between two or more robots that are considered to be in a peer-to-peer relationship is on an equal footing, and control functionality may fluctuate. For example, if we have two robots that we call Robot A and Robot B, and if they are in a peer-to-peer message-passing or control relationship, then Robot A sometimes may be the sender and sometimes the receiver. Or Robot B could start out as the initiator and become the consumer. Peer-to-peer coordination can be tricky using MINDSTORMS NXT because in normal Bluetooth communication conditions, only one robot can be considered the initiator, and that robot can direct messages up to three receiving robots. But the three receiving robots can only direct messages to the initiator and not to each other.

Robot in the Middle?

Let's bring our three robots back, Robot A, Robot B, and Robot C. For demonstration purposes, only Robot C is the initiator and connects to Robots A and B. Under normal conditions, Robot C can send messages to both Robot A and Robot B. Robot A can send messages to both Robot C and Robot B, and Robot B can send messages to Robot C. However, Robot A cannot send messages to Robot B, and Robot B cannot send messages to Robot A (at least not directly). So how can Robot B tell Robot A about a local external event or a local internal event? Well, for these versions 1 and 2 of the NXT controller, the most practical way is to use Robot C as a go-between. That is, if Robot A wants to send a message to Robot B, then Robot A must first send the message to Robot C, and then Robot C must pass it on to Robot B. This would be the case if Robot B wanted to communicate with Robot A. This "robot in the middle" approach is one way to simulate peer-to-peer coordination when it is not available directly. There is another way to accomplish peer-to-peer coordination if we're using events.

Peer-to-Peer Coordination Using Global External Events

We could use global external events to help coordinate things in a peer-to-peer team. Because two or more of the robots perceive the same external event, they can use that event to coordinate with each other without sending messages. Yes, yes, we know, it's not really peer-to-peer communication if no one is sending messages. But keep in mind that peer-to-peer communication is related both to the communication mechanism and the control mechanism. We will have much to say about this later, but if all the robots have some way to perceive the same global external event at the same time, and if perhaps one robot can see it, the other can hear it, and the other is touched by it, then no one robot is really in control of the information or the other robots, and they are, in effect, acting as peers. Of course, this is a little of a hack, but all is fair in love, war, and when you're trying to get all your robots to work together as a team. We use the global external events technique to make the case that there's always a way out or in. The "robot in the middle" approach is a crude way to accomplish peer-to-peer communication and control, but it does work. We also can emulate communication pipelines, client-server coordination, and producer-consumer coordination between any two robots using the "robot in the middle" approach as a simple pass-through.

The Team-Leader Approach: It's All About Perspective

We will take advantage often of the team-leader approach. In this scheme, one team member gets to play captain and control the information flow and dispatch the other team members to execute various tasks. In this scheme, the team leader's work is largely coordination, planning, interpretation, and decision making. The team leader can also be involved in heavy lifting as well in conjunction with team-leading duties. Here's where the fun comes in with our BRON approach to creating teams of robots. The choice of team leader can have a dramatic effect on how the robot application is designed and conceptualized. Your imagination and the possibilities will go in different directions depending on who the team leader is and the type of communication model you choose among the team members. For example, if the BRON is put together with the computer as the team leader, then are we designing a computer program where robots and other Bluetooth devices are simply peripherals under service of the computer program. If we have an Android phone, an NXT-based robot, and a computer, and we choose to make the robot the team leader, then are we really just using the Android phone and computer as resources for the robot to accomplish its task? What if the Android phone is designated the team leader? Are the computer and robot just extensions of the Android's text messaging and communication facilities? Or combined, do the computer, NXT-based robot, and Android phone form something unique? A totally new form of hybrid automation? Of course, the answer is all the above and much more. It's all a matter of perspective. Which device takes the lead is up to you. This means that your approach to the solution to a problem or execution of a task is only limited by your imagination and experience.

The MINDSTORMS NXT robotic kits offer an incredible range of possibilities for performing tasks and solving problems. When you use the Bluetooth communication protocol and capability to add computers that may range from notebook computers to supercomputers and smart mobile devices such as the Android phone, then the possibilities truly are limitless. If we further combine all these possibilities into one or more BRON teams working together, the only conclusion we are left with is that if we can't build it under these circumstances, then it can't be built!

The Message-Passing Approach to Coordination

The message-passing approach is perhaps the simplest, and at the same time, it has the potential to be the most complex while also being among the most powerful techniques for a coordinating a team of robots. Because this book is only an introduction to building and programming teams of robots and a mix of other Bluetooth-enabled devices, we will only scratch the surface of many of the techniques available for coordinating a collection of robots or agents. We will use simple producer-consumer coordination and team-leader and team-member coordination, but we will also throw in a few multiagent communication and control schemes so that you will have some idea of how vast the possibilities are for your robotic creations. At the bottom of most of these schemes will be message passing, using Bluetooth, connect, wait, read, write, monitor, and close capabilities.

The Basic BRON Approach

The basic BRON approach leverages the hard work and creativity that have gone into your existing robots by showing you how to put your robots into teams or how to add your robot to a team by taking advantage of the Bluetooth communication capabilities of the NXT microcontroller and by adding team-mode modules to your already existing robots or including team-mode modules in the design of any new robots that you will be building. The idea is that with little or no change, you can construct a team of robots/devices to do things collectively that individually they could not do. We approach with dread the mere idea of having to reprogram our robots or tear them down and start over just to get a little more functionality. So we are very sensitive to fellow roboticists who have similar fears and dreads. Of course, we will recommend adding a team mode when you are in the design phase of your robotic creation just in case your robot may be invited or drafted to work on some robotic team. This is the most flexible design approach. We'll show you object-oriented techniques and agent-oriented techniques to accomplish these kinds of robot creations. However, if your robot has already been built and programmed, then we will be recommending techniques that will attempt to minimize any change to existing code in order to take advantage of the teamwork techniques and the Bluetooth communications capabilities. Remember, every device that speaks a compatible dialect of Bluetooth is a potential team member in the BRON, and that device brings all its capabilities and strengths to a potential BRON team.

The World of Bluetooth Devices

Here's the trick we will use to expand the world of Bluetooth devices that our NXT robots have access to: We know that we can connect our NXT-controlled robots to computers, and we know that we can connect our NXT-controlled robots to Android phones, but there are many Bluetooth devices that are not compatible with NXT controllers. In many situations, we will use the computer and the Android phone as the middleman because there are devices that are compatible with the Android but are not compatible with the NXT controller. Can we use the Android as a middleman to take advantage of some of those devices? There are devices that are Bluetooth-compatible with the computer that are not compatible with the NXT controller. Can we use the computer as the middleman to get at the capabilities of some of the devices that are not compatible with our NXT-controlled robots? Well, of course we can! Table 1-4 shows some of most commonly found Bluetooth-enabled devices.

TABLE 1-4 Common Bluetooth-Enabled Devices

Bluetooth-Enabled Device	Manufacturer	Description
USB Dongle	Abe UB22S	The USB Dongle sold by LEGO for the MINDSTORMS Robot; recognized Wiimote and connected right away; had some difficulties getting it to communicate with GlovePie, but it works with standard Windows XP Bluetooth stack.
	Acer BT-700 USB 1.1	Works with the newest BlueSoleil version.
	Silicon Wave Exter USB Dongle	
	SiteCom USB Bluetooth Dongle	
	Cellink BTA-6030	Uses BlueSoleil.
USB Bluetooth Adapter	Advent USB Bluetooth Adapter, v2.0 + EDR, Class 1 (P/N ADE-C1EDR)	Toshiba Bluetooth stack.
	Belkin Bluetooth USB Adapter F8T001, v2	Uses BlueSoleil, WIDCOMM untested.
	Belkin Bluetooth USB Adapter F8T003, v2	Uses BlueSoleil, WIDCOMM untested.
	Trust Bluetooth 2.0 EDR USB Adapter BT-2200Tp	Uses both WIDCOMM and BlueSoleil.
	Trust Bluetooth 2.0 EDR USB Adapter BT-2305p	Uses Toshiba stack.
Bluetooth Modules	Dell TrueMobile Bluetooth Modules	Standard issue included with many Dell laptops; Latitude D820 has onboard module that works best with BlueSoleil.

In Chapter 2 we will take a deeper look at Bluetooth, how it is used with NXT-based robots and how to create NXT-G Bluetooth bricks and LabVIEW for NXT Bluetooth VI that can control teams of robots.

All Robots Are Machines, But Not All Machines Are Robots

It used to be clear what one meant by the word *robot*. But computers, software, electronics, and battery-operated gizmos have changed all that. We now have all kinds of machines that seem to operate on their own and, depending on who you talk to, might be considered robots. For example, the microwave oven virtually operates on its own. It's a machine, and it follows a program—that makes it a robot, right? What about the nice five-speed oscillating fan that switches or moves from left to right blowing nice cool air without human intervention? Surely that's a robot. What about the modern washer and dryer? They have software and are programmable, and once they start, they perform a sequence of tasks automatically without human intervention—they're robots, too, right? And we can't leave out all the remote-controlled battle bots that are included year after year in the annual battle-bot competitions, can we? These are robots because the competition says they are. The line between computerized machine or mechanism, remote-controlled machine, and robot is pretty blurred.

So what exactly is a robot? Well, at Ctest Laboratories, we use a list of traits or features that any robot must have in order to define what a robot is. The following is the list of traits we currently use. For a machine to be considered a robot, it must meet at least the seven following criteria:

- It must be a nonliving machine.
- Its reprogrammable behavior, actions, and control are the result of executing a programmed set of instructions that execute without the need for external intervention.
- It must be able to sense its external and internal environments in one or more ways through the use of its programming.
- It must be able to affect or otherwise interact with or operate on its external environment in one or more ways through the use of its programming.
- Once initiated, it must be able to execute its programming without the need for external intervention, and it must have some degree of mobility.
- It must have its own power source.
- It must have a language that is suitable for the representation of discrete instructions, and the language must support the notion of programming.

If a machine does not meet at least these seven criteria, then, for our purposes, it is not considered to be a robot!

Chapter 2

Bluetooth for MINDSTORMS NXT: A Closer Look

The Lost Scrolls of Robotics: #2

Can you see around corners? No, merely through walls.
— Robby, *The Invisible Boy*

We've already taken a sneak peek at how Bluetooth can be used to allow two or more NXT robots to communicate. Now let's take a closer look and delve a little deeper into exactly what Bluetooth is and how it can be used with NXT-based robots to promote the concept of a team of robots working toward a common goal or how to allow a group of NXT devices to solve a problem collectively that they could not solve individually.

So Exactly What Is Bluetooth?

Bluetooth is a wireless communications technology that is used to transfer or exchange data over short distances using short-range radio-wave transmissions. How short is short? Bluetooth operates in the 2.4- to 2.483-MHz range. This range just happens to be among the ranges used by many microwave ovens, cordless phones, and countless other short-range low-power communications systems. The official name of this range is the *Industrial, Scientific, and Medical* (ISM) radio band because it is internationally reserved for industrial and medical purposes. But its low power and short range happen to suit Bluetooth just fine. The power needed for Bluetooth is 2.5 to 3 mW. Bluetooth has an official range of about 10 m. At the time of this writing, Bluetooth has a practical data rate of about 1 Mb/s (theoretically, it is faster, but in practice it is 1 Mb/s), but the industry is pushing for a faster practical date rate of 2 or 3 Mb/s. The length of a Bluetooth wave is about 12.5 cm. Bluetooth is typically used to transfer or exchange data between mobile phones, personal data assistants (PDAs), printers, digital cameras, joysticks, gaming systems, desktop/notebook computers, computer peripherals (e.g., keyboards and mice), and robots. In addition to simply transferring data, the Bluetooth technology can be used as a control technology as well (i.e., the data sent could represent commands). The list of Bluetooth-enabled devices is constantly growing. One of the nice features of the radiofrequency (RF) wave that Bluetooth uses is that it can move through physical structures such as walls and floors. Yes, the signal can move in virtually all directions simultaneously. This means that as long as two devices are in range, they can be in different rooms or even on different floors. This is possible because of the physical shape of the waves. Figure 2-1 shows a dipole antenna model, the gain of a half-wave dipole, and the three-dimensional (3D) radiation pattern of the dipole antenna that can be taken on by Bluetooth communications.

For the most part, the shape of the wave is determined by the kind of antenna that transmits the Bluetooth signal. Different antennas are often used not necessarily for power, but to maximize or control the direction in which the Bluetooth signal radiates.

The Myth of NXT's Bluetooth Problem

Sometimes the NXT brick gets a bad rap for being restricted to a one to three master-slave type of Bluetooth configuration. In the NXT implementation of Bluetooth, a single master NXT brick can communicate with up to three slaves. Recall from Chapter 1 that there are several communication protocols that cannot be implemented directly using NXT and Bluetooth. A peer-to-peer network is one example of a communication scheme that is not directly supported because of the master-slave configuration. This is where the NXT myth comes from. In fact, the master-slave structure is actually a feature of (or problem with) the Bluetooth protocol/technology itself, which is typically implemented as a master-slave architecture with one master device able to communicate with up to seven slave devices. Yes, it is true that NXT is limited to no more than three slave devices, but the overall structure comes from the Bluetooth standard.

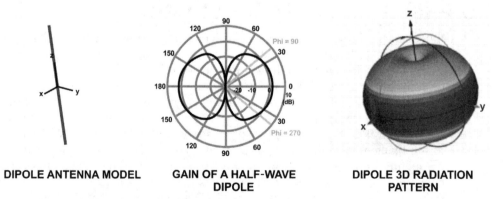

DIPOLE ANTENNA MODEL   GAIN OF A HALF-WAVE DIPOLE   DIPOLE 3D RADIATION PATTERN

FIGURE 2-1   Dipole antenna model, gain of a half-wave dipole, and the 3D radiation pattern of the dipole antenna.

What Does Bluetooth Mean for NXT-Based Robots?

The more Bluetooth-enabled devices there are out there that are compatible with NXT, the more flexibility our robots will have. The Bluetooth technology can be used for personal area networks (PANs) of NXT robots coupled with other Bluetooth-enabled devices. The Bluetooth PAN is also referred to as a *piconet*. In contrast to most wide area networks (WANs) and local area networks (LANs), PANs are transient, ad hoc, or even spontaneous. That is, they are not normally permanent or even semipermanent networks. Two or more devices hook up for some unspecified time, disconnect, and that's the end of the PAN or piconet for the time being. PANs or piconets typically are not permanent networks. Although the participants in the network may store the connection information permanently, the network itself is usually transient. Thus, thanks to Bluetooth, we can have an NXT-robotic PAN (or piconet) that can be used as the basis of communication and control between a team of NXT-based robots and other Bluetooth-enabled devices. If a Bluetooth-enabled device is compatible with NXT, then we can add the capabilities and features of that device to our team of robots through communication. Recall that in most cases, effective teamwork requires effective communication. Because the NXT brick has Bluetooth technology built in, this means that NXT-based robots can join and participate in Bluetooth-based PANs, each Bluetooth-enabled device bringing a potentially unique set of capabilities to the piconet. Each device is able to communicate either directly or indirectly so that capabilities can be shared, directed, or controlled. Thus, when we think about tasks that we want our team of robots to execute or problems that we want them to solve, we can now consider the basic building block, the NXT robots, as well as any compatible Bluetooth device to which we have access.

Is NXT-Bluetooth Capability Software or Hardware?

A Bluetooth-enabled device in all but the most exotic of cases will require a hardware component (which includes an antenna) and a software component that implements the Bluetooth stack or Bluetooth protocol. NXT has an onboard microcontroller that helps to deliver Bluetooth, and it has firmware and an antenna. Figure 2-2 shows a logical block diagram of the NXT brick.

Notice that the Bluetooth chip module is at the top of the NXT brick relative to the ARM7 microprocessor and the Atmel coprocessor. The figure also shows a simplified but exploded view of the Bluetooth chip module. It contains three basic components. Note that the rightmost component on the Bluetooth chip module is the antenna. This is the antenna that will be largely responsible for the shape and strength of the signal that the NXT will send. Although adding an external Bluetooth antenna is possible (which might boost range and direction), we have found that the built-in antenna can and does send, for all intents and purposes, an omnidirectional signal that moves past

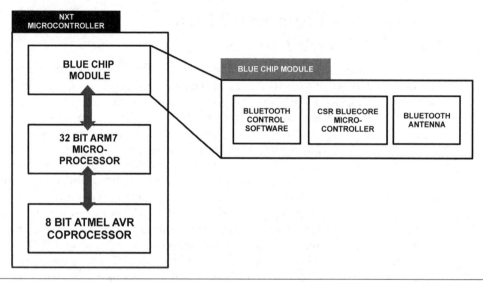

FIGURE 2-2 A logical block diagram of the NXT brick.

and through walls and over multiple floors (sometimes as far as 12 m) of the buildings where we run our robots. The Bluetooth control software that is shown in Figure 2-2 is the lowest layer of the Bluetooth software. We interact with that software through software that is found at higher layers in the system. Figure 2-3 shows a simplified overview of NXT software layers.

We typically only interact with the Bluetooth NXT-G/LabVIEW blocks, Java class libraries for Bluetooth available in the leJOS environment, or the C function libraries in the NXC or ROBOTC languages. Unless we're prepared to change the Bluetooth chip module shown in Figure 2-2, our Bluetooth options are limited to what's available at Layer 3, as shown in Figure 2-3. In this chapter we focus primarily on Layer 3 as delivered in the LabVIEW for MINDSTORMS toolkit and the NXT-G software that comes as the default language with the MINDSTORMS robot kits. We explain how to use C and Java Bluetooth functionality in Chapters 5 and 6, where we take a closer look at programming teams of robots using NXC and the leJOS environments.

A Pause for Some Bluetooth–NXT Brick Preliminaries

Two features that aren't so important when building a single robot but that become relatively important when that robot participates with other robots or devices in a team setting are

- The NXT brick's name
- The personal identification number (PIN)

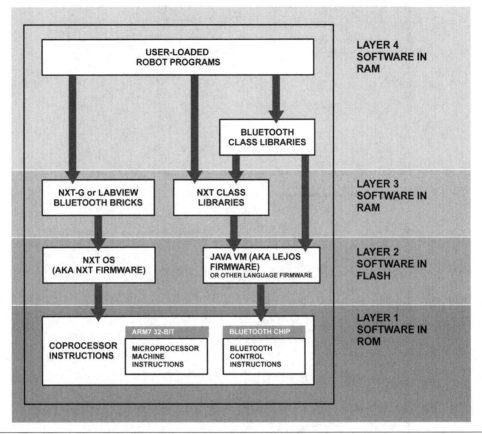

FIGURE 2-3 A simplified overview of NXT software layers.

What's in a Name?

As you will see, one of the first steps in setting up a Bluetooth PAN is to scan for all the devices that will participate in the network. The default name for every MINDSTORMS NXT brick as it is delivered from the factory is "NXT." Thus, if you have several NXT robots that will be involved in the network and you run the Bluetooth scan, they will all show up as NXT. The first thing we recommend is that you give each NXT brick a name. As you may recall from Chapter 1, we give all our robots meaningful names that identify what team each robot is a part of and what stage of capability each robot has. For example, our A3R3 robot is a member of team A and is a stage 3 (fairly advanced) robot. It is a robot (R as opposed to D for a device) and is the third member. Obviously, you can choose any name you like for your robot, but remember, whatever name you choose will be what shows up on the list during any Bluetooth scan that is in range of your robot. We do recommend that you give every device participating in the team a unique name. Figure 2-4 is a screen capture of the LabVIEW and NXT-G locations where the name of the NXT brick is changed.

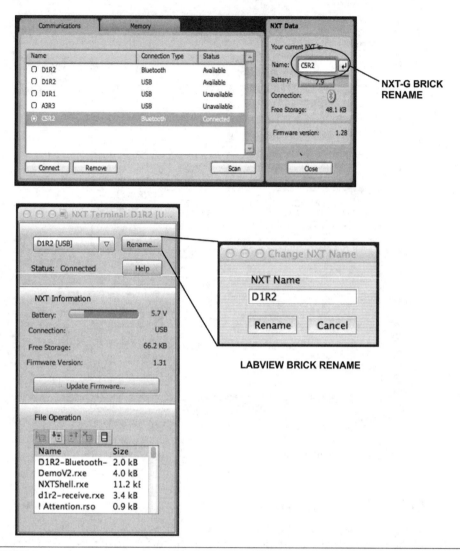

FIGURE 2-4 A screen capture of LabVIEW and NXT-G locations where NXT brick names are changed.

The same holds for computers, Android phones, or other Bluetooth-enabled devices. In some cases, the name will be unique because it will default to some kind of address for the device. In other cases it just defaults to the model name of the device. In this book we make sure that each device that participates in a team has a name that reflects its function on that team. For example, for our Linux computers, we used the .conf files and Kbluetooth, as shown in Figure 2-5, to set the name of our device.

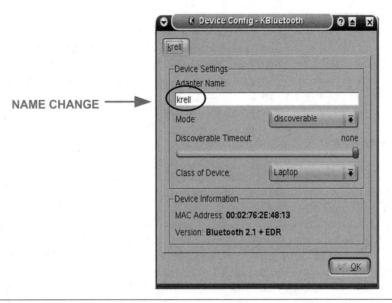

NAME CHANGE

FIGURE 2-5   A screenshot of the device name change in Kbluetooth.

A Little Security (or at Least Privacy), Please!

Every NXT brick has a four-character PIN. This PIN is required by the receiver when the initiator tries to pair up or make a connection (more on initiator and receiver later on). By default, the PIN is "1234." Again, we suggest that you devise a PIN that is meaningful in the context of your team. This is just a suggestion, but do change it to something. As Bluetooth becomes more pervasive, there will be little piconets all over the place. You don't want to have your piconet party invaded by an unwanted guest. The Bluetooth signal is stronger than you think, and 10 to 12 m farther might make your piconet visible in places you don't expect it to be visible. A PIN of 1234 is not very hard to guess. Figure 2-6 is a screenshot of the PIN change option on the NXT brick with leJOS JVM.

From BLUETOOTH submenus   **Choose "Change PIN" option**   **Enter your new PIN number**

FIGURE 2-6   PIN change option on the NXT brick with leJOS JVM.

Visibility vs. Invisibility

Before two or more Bluetooth-enabled devices can communicate, they must be aware of each other. The Bluetooth devices that we have as potential team members will all have a scan ability. The scan ability allows the device to search for other Bluetooth-enabled devices that are

- Within range [range $\times$ 10 m (give or take a meter here or there)]
- Visible

In order for the scan feature to pick up a potential partner, the potential partner must be in range and have visibility or discoverability turned on. A device may be in range and have visibility or discoverability turned off, which will cause it not to appear on your scan list. Figure 2-7 shows the NXT scan function and what a scan list looks like when it discovers other visible Bluetooth devices that are in range.

When a device appears on the scan list, you select it and try to pair with it. If you know the correct PIN, the NXT brick will give you the "connected" message. Once you are paired with a device or connected with an NXT brick, you no longer have to scan for it. Paired devices save the addresses and names of the devices with which they have connected successfully. Once you have established all your team members, then you can turn visibility off to prevent your devices from being listed on other scans.

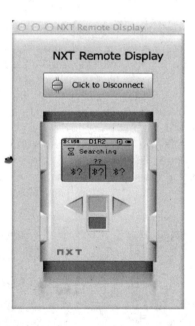

SCANNING DEVICES

VISIBLE BLUETOOTH DEVICES

FIGURE 2-7   The NXT scan function and what a scan list looks like when it discovers other visible Bluetooth devices.

Devices that are already paired with each other will be able to connect using the address they have on file. Although turning visibility off does not completely shield your device from connection attempts,* it's a good start. Changing your PIN and only having your device visible during the initial pairing sessions are good practices for establishing a little privacy and security.

Manually or Programmatically Connecting NXT Bricks

The Bluetooth connection between NXT bricks also can be accomplished automatically through programming. Figure 2-8 shows the LabVIEW and NXT-G blocks for connecting two NXT bricks.

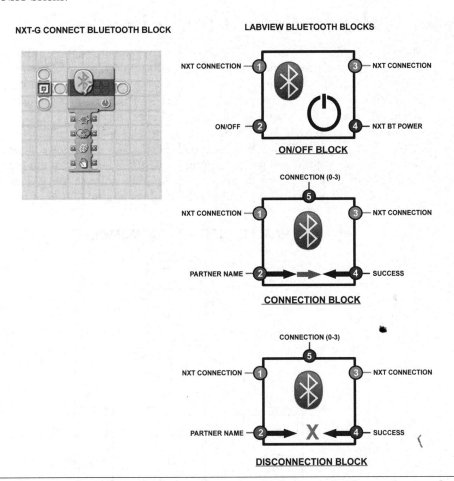

FIGURE 2-8   LabVIEW and NXT-G environments for turning Bluetooth on and for connecting and disconnecting.

---

*If a device can guess the address of your NXT brick, it can connect directly without the scanning process regardless of the visibility settings!

LabVIEW has a block that turns the Bluetooth feature on. It has another block that makes the connection between the brick running the program and a named brick. Notice the parameter in Figure 2-8 that captures the name of the partner brick. It also has a block for disconnecting bricks. The figure shows a similar block for the NXT-G environment. These blocks also allow you to turn on the Bluetooth capability, initiate a connection, and disconnect all from within a program instead of manually turning on Bluetooth or manually connecting/disconnecting the bricks. The NXC and Java environments have similar functions that we will cover later that will allow you to bypass the manual procedure of connecting two NXT bricks.

Who Is the Initiator (Team Leader)?

When setting up connections, the device that does the original scanning and contacting of the other devices is often referred to as the *initiator* or *master device*. Throughout this book, we use *team leader* to identify the device that makes the initial connection. If our team consists of only NXT robots, then the team leader will perform the scan, and it will hopefully find a list of willing participants. As the team leader selects each NXT device, it will be assigned to line 1, line 2, or line 3. Line 0 is reserved for the team leader. When the team leader wants to send a message to the devices it has added, it will use the appropriate line. When any of the team members want to send a message back to the team leader, they will use line 0. Figure 2-9 shows the classic team leader–team member relationships among four NXT bricks.

TEAM LEADER – TEAM MEMBERS

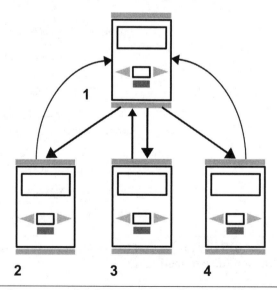

FIGURE 2-9   The classic team leader–team member relationships among four NXT bricks.

Notice that the team leader connects to three bricks, but none of the bricks connect to each other. Thus, if brick 1 wants to talk to brick 3, then brick 1 must send its message to the team leader, and the team leader sends the message on to brick 3. This has important implications for the architectures that are supported by our PANs or piconets.

Physical Architecture vs. Logical Architectures

Although the NXT bricks are only connected in a physical architecture of initiator–receiver or team leader–team member, we can emulate other team-oriented architectures logically, such a pipeline or peer to peer. The physical architecture reflects the actual connections of the NXT bricks. The logical architecture is how the communications and coordination are actually used among the bricks. At the programming level, one can emulate many different configurations, although at the connection level the bricks are in an initiator-receiver or master-slave configuration.

After the Connection Is Made

Turning Bluetooth on, setting visibility off/on, scanning, and making a device discoverable/not discoverable are pretty much the same for most of the devices that are compatible with the NXT brick: The initiator scans, finds devices, and makes the connection. With the Android phone, you also will be able to initiate, scan, and then list the NXT bricks as one of the types of devices to which it can connect. If the NXT does not have a name change, then "NXT" will show up twice on a list. And just the string "NXT" doesn't give you additional information about the nature of the robots listed. Again, this is why we suggest naming your NXT bricks right off the bat. Once the connection is made, there really are only two basic commands that are used between two or more bricks:

- `WriteMessage`
- `SendMessage`

or some variation (depending on the language used), for instance,

- `Read`
- `Write`

The message can be data, numbers, strings, or Boolean variables. The meaning of the message is application-dependent and can be whatever you want it to be. For example, one could send the number 4 to two pairs of robots. For one pair, the 4 is interpreted as a color that is used by the robots. For the other pair, the 4 can stand for

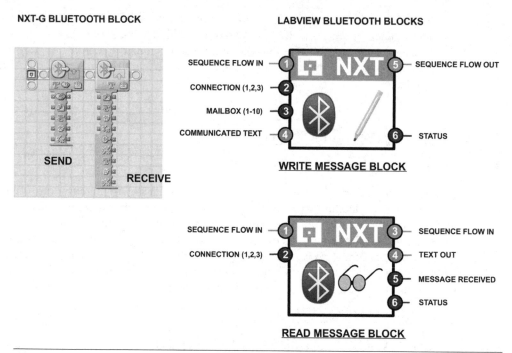

FIGURE 2-10   The send/receive and read/write bricks for the NXT-G and LabVIEW environments.

how many more seconds you want the robots to move forward. Figure 2-10 shows the send/receive, read/write bricks for the NXT-G environment, as well as the LabVIEW for NXT environment.

The message meaning is completely up to you. Because there are primarily only two basic functions involved with Bluetooth communications, we have to ask ourselves, how bad can it be? Well, it's relatively easy if you keep in mind the following points:

- Bluetooth functions do not automatically wait.
- Initiators receive messages only on line 0.
- Receivers receive messages from an unspecified line (understood as line 0).
- Receivers send messages on line 0.

Bluetooth Functions Don't Wait

Many communication protocols have the notion of reading a message and writing a message. Bluetooth is no exception. The message is typically some kind of number or string. Many of the protocols require the reading/writing of the message block until the receiver has completed the process. The blocking prevents the next command from executing until the message is read by the receiver or written by the sender. For example, if the Bluetooth messages were blocking:

```
ReadMessage(Distance)
Motor.forward(Distance)
```

The command `Motor.forward(Distance)` will not execute until `ReadMessage(Distance)` actually has read the information from the sending robot. In this case, `ReadMessage` would be considered a *blocking* call. However, what we have in the basic calls to Bluetooth for NXT are *nonblocking* calls. This means that the `ReadMessage` command would execute immediately and then try to move the motor forward, whether it had received any information concerning distance or not. In the case of the NXT-G blocks for read and display in Figure 2-11, even if the string to be displayed has not been received, the display command would execute. This nonblocking behavior is probably the biggest headache for Bluetooth communications on the NXT robot. Thus, to make the communications work, we have to create our own blocking (waiting) logic. This usually involves using a wait block or a time block. It also can be accomplished by using a while loop that waits on some condition or a for loop that executes a particular amount of times. The whole idea is to have flow in the program or wait until the message has been either sent or received.

Talk to Initiators on Line 0

If you are using NXT-G or LabVIEW and the initiator is an NXT brick, then receivers must send any message intended for the initiator on Line 0. Recall that lines 1

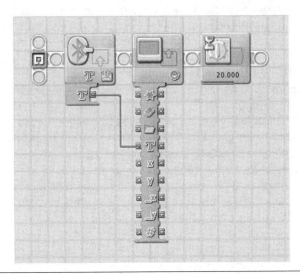

FIGURE 2-11   NXT-G program read and then displays a Bluetooth message.

through 3 are used by the initiator to send messages to receivers or team members. That is, the initiator will talk on lines 1 through 3 and listen on line 0. Because all team members in this configuration send information back to the initiator on line 0, in most cases it is a good idea for each team member to somehow send its identity along with its message. For example, the team robots that we build might send to the initiator string formatted:

```
"D1R2 liquid is harmless"
```

This string lets the initiator know that robot 2, stage 1, from D sent the message that the liquid is harmless. In this way, all members can send messages back on line 0, and the team leader will know where the messages came from.

Introducing the Scout Bots

Over the next several chapters we will begin to construct our crime scene investigation (CSI) team of robots. The CSI team will be designed to do simple crime scene investigation and analysis. Well, in reality, it is simple analysis of a *simulated crime scene*, and to be more exact, it is a simulated *staged crime scene*. The reader will be taken on a virtual crime scene investigation in which the names and incidents have been changed to protect the innocent. Ultimately, we will use the CSI team (team D) and one of the other teams we will introduce in this book to deal with a compromised warehouse. For now, though, we will introduce you to D1R1 and D1R2. These are the primary scout bots that we will use for examples in the next several chapters. The capability matrix for D1R1 and D1R2 is shown in Table 2-1.

Recall the naming convention. D represents the team. D1 means that this is a first-stage robot. The fact that it is a first-stage robot means that it is a work in progress and has not been completed yet. The R1 and R2 represent the fact that we are working with Robot 1 and Robot 2 of the D team. Eventually, we will add a laptop computer to the D team, and the designation will be D3C1 (C for computer). In stage 1, D1R1 does 16-color spectrum sensing and ultrasonic sensing and has an arm and a gripper for an effector. D1R1 can retrieve and place relatively large objects for its size. Figure 2-12 shows the D1R1.

D1R1 is built using the basic MINDSTORMS NXT kit, a HiTechnic color sensor, and the Tetrix base robot kit. D1R1 is the first robot in a three-device team. The second robot is D1R2. D1R2 is used primarily for chemical analysis and uses a pH probe sensor from Vernier.* Figure 2-13 shows the D1R2.

Our team struggled with the decision between exotic designs, perhaps unique robot designs, or off-the-shelf robot designs for this book. We currently have several interesting teams of robots that we use at Ctest Laboratories that we considered as

---

*Vernier is a sensor manufacturer that makes many awesome probes and sensors that are compatible with MINDSTORMS NXT robot kits. There are also LabVIEW and NXT-G blocks for all of the Vernier sensors we use in this book.

TABLE 2-1 The Capability Matrix for D1R1 and D1R2

Team (Robots/ Devices)	Microcontroller/ Microprocessor	End Effectors	Mobility Type	Display	Sensors/ Motors/ Controllers	I/O Ports	Ranges/ Degrees	Communication
D1R1	32-bit ARM7, 64-kB RAM	Arm and Gripper	4 Wheels	100- × 64-pixel liquid- crystal display (LCD) screen	LEGO touch	3	1–0	Bluetooth, USB
					HiTechnic color	2	16 colors	
					LEGO ultrasonic	4	0–255 cm with a margin of ±3 cm	
					Direct-current (DC) motor controller	1		
					Left wheel DC motor	DC controller 1– 1+	360 degrees	
					Right wheel dc motor	Dc controller 2– 2+	360 degrees	
					Left wheel encoder	Enc. 1	360 degrees	
					Right wheel encoder	Enc. 2	360 degrees	
					Arm servo	Servo controller port 1	90 degrees	
					Gripper servo	Servo controller port 2	90 degrees	
D1R2	32-bit ARM7, 64-kB RAM	Probe and touch sensor extension	2 tractor wheels	100- × 64-pixel LCD screen	LEGO touch	1	1–0	Bluetooth, USB
					NXT adapter/ pH sensor	3	pH 1–14	
					Probe servo	B	90 degrees	
					Right tractor wheel servo	C	360 degrees	
					Left tractor wheel servo	A	360 degrees	

FIGURE 2-12   Photograph of D1R1.

potential centerpieces for this book, but we decided to keep the focus on building and in some instances repurposing teams of robots using Bluetooth by using the base bots or starter-type bots that come with most robot kits today. While we do stretch out on the robot designs a little, most of the examples in this book use basic robot designs (with few enhancements or changes) that should be familiar to builders who use the MINDSTORMS NXT and Tetrix kits. Although we don't explicitly use the Vex robot kit, many of the base robot designs that we use (e.g., Tetrix Ranger Bot base) are very similar to the basic Vex starter kit. And the robot teamwork concepts that we introduce apply equally.* Therefore, rather than focusing on several complex robot builds, we decided to keep designs simple and accessible because communication with and coordination of two or more robots are challenge enough. We did not want to add to that the use of unfamiliar robot designs. However, we do modify the designs of all base robots so that they can accomplish a little more than what they were designed to do. In some cases, we made enhancements such as adding new end effectors or

---

*Although most likely you would replace Bluetooth with VexNet as the communications vehicle.

FIGURE 2-13   Photograph of D1R2.

robot arms. In other cases, we swapped out stock LEGO sensors for HiTechnic or Vernier sensors that had more capability. One of our guiding principles in writing this book is to give readers tools, tips, techniques, and processes that will allow them to get more out of the robot(s) they currently have.

D1R1 has extensions of the base robot kits that come with Tetrix. D1R1 is based on the famous Tetrix Ranger Bot. The primary change that we made to the Ranger Bot was to add a robot arm with a gripper as an end effector. Although this is a common extension to make to the Ranger Bot—and in fact, there are standard robot build instructions for an arm—we chose to modify the standard extension and place our robot arm on the left side of the robot, bring it a little lower, and make the support beam shorter. While D1R2 is a standard LEGO Tractor Wheel design, we added an extension to the back of the robot in order to manipulate the probe and position the touch sensor. The touch sensor is used to detect a beaker or container into which the probe will be lowered to perform chemical analysis. The probe shown in Figure 2-13 is a pH sensor. In several of our examples, D1R1 will discover a substance or liquid that needs to be identified in some way. D1R1 typically will deliver the substance to D1R2 to do the analysis. D1R2 will work together with D3C1 to come to a final conclusion about the substance or liquid and whether it is dangerous or not. The D team communicates through Bluetooth. To give you some idea of what this looks like in LabVIEW, we will set up D1R2 to be the team leader and D1R1 to be a team member. The pH sensor does not appear in the schematic.

Setting Up the Initial Bluetooth Connection

Figure 2-14 shows the basic steps for setting up the initiator or, in our case, the team leader. In Figure 2-14*a*, the Search submenu on the Bluetooth menu is selected on the D1R2's NXT microcontroller. Once we have the menu, Figure 2-14*b* shows what the Bluetooth search option looks like. Figure 2-14*c* shows a list of Bluetooth-enabled devices that D1R2 to which may connect. This screen may be different for you

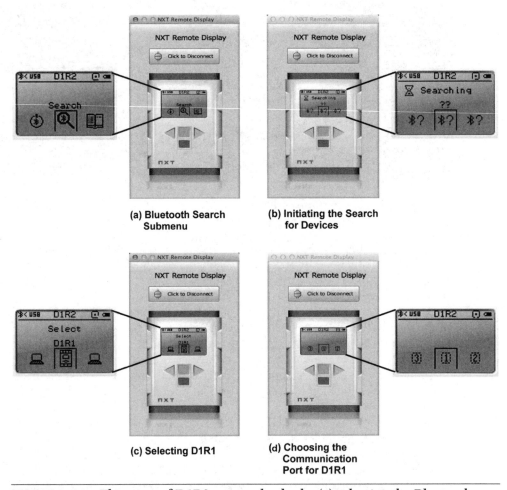

(a) Bluetooth Search Submenu

(b) Initiating the Search for Devices

(c) Selecting D1R1

(d) Choosing the Communication Port for D1R1

FIGURE 2-14   The setup of D1R2 as team leader by (*a*) selecting the Bluetooth Search submenu, (*b*) initiating the search, (*c*) selecting D1R1 as a team member, and then (*d*) choosing the communications port that will be used to communicate with D1R1.

depending on what Bluetooth-enabled devices are on and within range (~10 m) of your NXT brick. In our case, we had a couple of Mac powerbooks and D1R1, so we select D1R1. Figure 2-14*d* shows the choice of three lines to assign D1R1 to. This is important to remember. Whichever line or channel you select here will be used by the team leader to communicate with that particular robot. In our case, we select line 1 for D1R1. This means that when D1R2 wants to send a message to D1R1, it will do so over line 1. Notice in Figure 2-14*d* that there are only three lines: 1, 2, and 3. Here is where our architecture limit comes in. When we refer to the fact that the NXT has a master-slave configuration for Bluetooth with one master and three potential slave devices, this is where you will run into that limitation. Once you've identified a device with which to connect, you will only be given a three-line choice to connect the device to. There is no line 4. This is in contrast to the Bluetooth standard, which has a 7 to 1 ratio. That is, the standard allows for one initiator with seven potential receiver devices. Now, from the D1R2 perspective, D1R1 is connected to line 1. Information leaving D1R2 and going to D1R1 will go over line 1. D1R1 will receive information from the initiator or team leader on line 1. However, if D1R1 wants to send information to D1R2, it will need to use line 0. Team members always use line 0 to talk to the team leader.

Waiting for and Sending a Bluetooth Response

In this chapter we show the basic wait/receive Bluetooth protocol using LabVIEW for MINDSTORMS NXT. We will review this technique for Java and C in later chapters. Keep in mind that the coding for all the examples and most of the major projects can be found in LabVIEW, leJOS (Java), NXT-G, and NXC on the website www.robotteams. org. We use LabVIEW here to illustrate a few important techniques. Some robot builders will prefer LabVIEW and already know its benefits regarding sensor testing, motor testing, and NXT brick testing. Other NXT robot builders will prefer NXC, leJOS (Java), or NXT-G. We recommend using LabVIEW (if you have access to it) in the early stages of building your robots because LabVIEW can be used to

- Easily verify sensor ports
- Easily verify controller ports
- Visually check basic sensor functions
- Check motor functionality with no programming necessary

The LabVIEW Schematic Editor

We'll have much to say about this later, but for now we'll just mention the fact that early in the robot build you may get wiring wrong, you may connect a sensor to a port other than the one that it was intended for, you may have the servo channels mixed up, or maybe you are unsure of what output a particular sensor will have. Figure 2-15 is a screenshot of the schematic editor for D1R1. Notice that D1R1 has two direct-current (DC) motors, two servos, an ultrasonic sensor, and a touch sensor. Using this

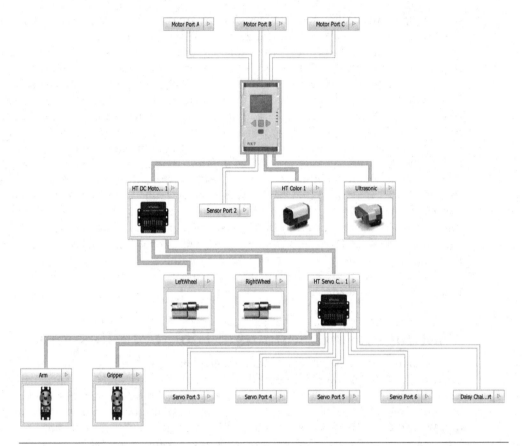

FIGURE 2-15   A LabVIEW schematic of D1R1.

schematic and zero programming, LabVIEW allows you to check ports, channels, and sensor outputs, and motor/servo assignments, among other things.

Teamwork: A Simple Bluetooth LabVIEW Application

In this simple application, D1R1 and D1R2 work together to determine whether a substance that has been presented requires analysis or not, and if it does require analysis, then the analysis takes place. The application is divided into two LabVIEW programs. One is executed by the team leader, and the other is executed by a team member. Both programs require Bluetooth communication blocks. This application demonstrates the basic steps that will take place for almost any Bluetooth session between members and represents the most fundamental step in robot-to-robot teamwork: sending and receiving messages!

The Team Leader Program (D1R2)

In step 1 in Figure 2-16, the team leader executes a `Wait For` communication block. This block takes a mailbox number as input. Note here that we didn't specify that the `Wait For` communication block requires the line number across which the team leader is communicating. The team leader is waiting for a message from a team member. All regular team members communicate across line 0 to the team leader. We don't have to worry about the line number of the team leader when programming the `Wait For` communication block. All we need is the number of the mailbox in which the message was placed. In this case, the team member (D1R1) places a message in mailbox 1 and sends the message on line 0. The important thing here is that we used the `Wait For` communication block. This causes the team leader not to do any further processing until it gets the message. It is tempting to use the Bluetooth `ReceiveMessage` function here, but that function does not automatically block or wait on the message. If we were to replace the `Wait For` communication block with the Bluetooth `ReadMessage` block, then the program in Figure 2-16 would not work correctly. Additional programming would be needed to create a *blocking* or *waiting* condition for the Bluetooth `ReadMessage` function. In step 2 in Figure 2-16, the team

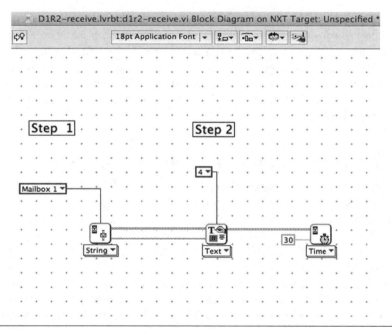

FIGURE 2-16   A simple LabVIEW program that is executed by the team leader.

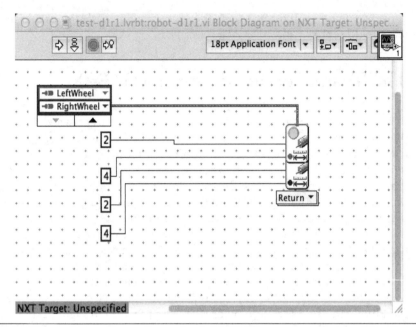

FIGURE 2-17 A short excerpt from the LabVIEW program executed by the receiver/team member (D1R1).

leader displays the message that is sent from team member D1R1 on the NXT brick display. Step 3 causes the message to remain on the brick screen for 30 seconds. If the message states that the substance is a solid, the team leader will relay that fact to D1C3. If the message is that the substance is a certain color liquid, then the team leader will request that the liquid be presented for analysis. Figure 2-17 shows the initial communication on the side of the team member that is doing a preliminary substance analysis.

The Team Member Program (D1R1)

Notice the Bluetooth `WriteMessage` box selects line 0 for the message. Keep in mind that D1R1 is a receiver and can only write messages to or send messages on line 0, whereas the initiator would have to select line 1, 2, or 3. The receiver must use line 0. The question is how long does the initiator have to wait. Well, from Figure 2-17, we can see that the entire operation will depend on getting a reading from a color sensor. Thus, at the very least, team leader D1R2 will have to wait on the "event" of reading the color sensor. Recall from Chapter 1 that one of the primary techniques for coordinating two or more robots is event coordination. Well, this simple program uses events to coordinate D1R1 and D1R2. The event that needs to happen is the sensing of a color. Once that color is sensed, the result is then sent to D1R2. Prior to receiving the color message from D1R1, D1R2 is in a hold pattern inside a `Wait`

`For` communication block. While this is a simple application, it does have the major components for a successful Bluetooth communication session:

- Initiator
- Send/write message function
- Blocking/waiting receive function
- Sender
- Receiver
- Message

Team Mode and Bluetooth in LabVIEW

Recall from Chapter 1 that we have a team mode attribute for our robot programs and a team mode Boolean variable that helped to determine whether the robot would be used in a standalone mode or as a component of a team. This same thing can be accomplished in LabVIEW using Boolean variables, `Wait` blocks, or `While` structures. Simply test the Boolean variable to see whether it is set to TRUE or FALSE to determine whether to execute code that has been placed in a team mode Sub VI or a team mode code branch. Typically, a Bluetooth application that requires robot teamwork will use no more than four or five Bluetooth blocks. Typically, there are Bluetooth blocks for

- Sending a message
- Receiving a message
- Waiting on the received message
- Shutting down communication

The `Receive` and `Send` blocks will be used the most, and the `Wait` block programming (which causes synchronization between two or more robots) will require the most attention in getting it right. Because the wait programming can depend on how many robots are involved, how many messages are sent back and forth, and the nature of the messages sent back and forth (e.g., string, number, or Boolean/logic), making sure that there are no errors in the waiting, sending, or receiving of the messages is the most challenging aspect of Bluetooth programming regardless of whether you are using LabVIEW, NXT-G, Java, or C.

In this chapter we've looked a little closer at Bluetooth. We've seen what it is, how it works, what its potential is, and how it can be used to communicate between two robots. But this is only the tip of the proverbial iceberg. We will have more discussion on Bluetooth blocks for NXT-G and LabVIEW in Chapter 5. We also will have a detailed discussion of the Bluetooth classes for Java and the Bluetooth functions for NXC. In our completed robot projects we will discuss how Bluetooth actually was used to accomplish the teamwork for each project. In Chapter 3 we will take a look at one of the major motivations for using more than one robot to solve a problem. Sensors! Not every robot can have every sensor needed. So why not combine robots through Bluetooth to take advantage of their sensor capabilities?

Chapter 3

One for All and All for One

The Lost Scrolls of Robotics: #3

I compute it to be an iconic directional probe searching for receiving outlets.

—Robot, *Lost in Space*

When designing your team of NXT robots, you have to decide what the robots will coordinate to do. What will be their task? Will they perform an experiment, locate an object, monitor a room, sort objects, or solve a Rubik's Cube or some other kind of puzzle? You have to determine what kind of robots you will need. Do they have to be mobile? What will each robot be capable of doing? Once you determine the tasks, break them down into subtasks, and then consider the robots on the team and their inventory of actuators and sensors. Whether your task is simple or sophisticated, it is a good start to be very informed as to what sensors and actuators can execute the task.

A single NXT brick has four input ports and three output ports. This means that you can connect the brick to at most four sensors, unless you use multiplexers, and three servo motors. LEGO MINDSTORMS manufactures six basic sensors, but there are over 100 sensors made by third-party vendors. These sensors can measure magnetic fields and the amount of pressure in a pneumatic system, the distance to an ultraviolet source, or the location of a teammate. To create an effective team of robots that are as autonomous as possible, it is important to know the range of sensors and actuators available and to understand how they work and their capabilities.

What Are Sensors?

Sensors and actuators are devices that make the robot an interesting piece of machinery. They play a big part in determining the extent of the autonomy of the robot. For example, a robot with motors and no sensors would move forward, running into obstacles, and continue to try to move forward until its energy source was depleted or somebody helped it. But a robot with a touch sensor would bump into something and maybe back up and try to navigate around it. Or it could use a scanning sensor and navigate away from obstacles that are within a predetermined range. In either case, a robot with the help of sensors could move around an area without help from people.

A robot can be equipped with many types of sensors and actuators that will allow the robot to engage in all kinds of physical behaviors. Some robots have human counterparts, such as sight, touch, and mobility, whereas others have animal counterparts, such as a bat's ability to locate an object using sonar. Nevertheless, robots can simulate these abilities by using sensors or combining sensors with actuators and devices. People have more than five senses, more like nine, but some researchers expand this list to 20. These include the senses of

- Tension
- Equilibrioception
- Thermoception

to name a few. Can you think of the sensors a robot would use to simulate these human senses? Table 3-1 lists some senses as human, robotic, or shared by both. It may be necessary to combine several sensors and actuators to achieve each of these senses. Robots can be equipped with thousands of sensors and still only be able to engage in some pretty narrow activities.

TABLE 3-1   Human and Robot Senses

Sense	Human	Robotic
Sight	Eyes (color receptors, rods for brightness)	Camera, light/color detector
Taste	Taste receptors	pH, colorimeter, dissolved oxygen sensor, orp, salinity, turbidity
Smell	Smell receptors	?
Tactile	Nerve endings	Touch sensor, artificial skin
Sound	Ear drums	Sound sensor and speakers
Kinesthesioception (movement and acceleration)	Utilizes nervous system	Accelerometer
Equilibrioception (balance)	Inner ears (vestibular labyrinthine system)	Gyroscope
Tension	Muscles	Combined strain gauge and force sensors
Thermoception (heat)	Hot/cold receptors	Barometer, temperature sensor, infrared thermometer
Magnetoreception	N/A	Magnetic sensor

TABLE 3-1   Human and Robot Senses (*continued*)

Sense	Human	Robotic
Time	Cerebral cortex, cerebellum, and basal ganglia	Clock
Hunger	Ghrelin hormone	N/A
Thirst	Thirst receptors	N/A
Echolocation (navigation)	N/A	Ultrasonic sensor, compass, GPS sensor
Electroreception (electric fields)	N/A	Electric-field (EF) proximity sensor
Direction	Hippocampus and entorhinal cortex (EC)	Compass sensor
Proximity	N/A	Ultrasonic sensor, EOPD, infrared sensor
Force, pressure	N/A	Force sensor, pressure sensor

Sensors: The Input Transducers

Sensors are transducers—devices that convert one form of energy to another. Transducers are used to sense different forms of energy, such as:

- Movement
- Force
- Electrical signals
- Radiant energy
- Thermal energy
- Magnetic energy

Transducer types can be used for input or output conversions depending on the type of signal or process being sensed or manipulated. Sensors, which are *input devices*, change a physical quantity into its corresponding electrical signal that can be mapped to a measurement. The physical quantity is normally nonelectrical. For example, the NXT sound sensor is a dynamic microphone. It detects acoustic waves and then changes them into an electrical analog signal. The physical quantity in this example is the sound wave. The electrical signal is generated by a diaphragm, a thin sheet of metal, that collects sound waves, causing the magnets that surround the diaphragm to vibrate. The vibration of the magnets causes a metal wire coil that surrounds the magnet to vibrate as well. This induces a current in the coil that is converted to the electrical signal. Then the signal is measured, indicating how loud or how soft the wave is. It is measured in decibels or adjusted decibels, which reflect the sound pressure detected. Thus the sensor will produce a signal that is proportional to the measured quantity. Actuators, which are output devices, change an electrical signal into another form of energy. For example, a loudspeaker changes an electrical

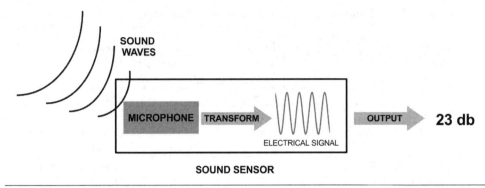

FIGURE 3-1   The conversion of sound waves into an electrical signal measured in decibels.

signal into sound waves. Figure 3-1 shows a sensor changing sound waves into an electrical signal that is measured in decibels.

Sensor Types

There are so many different types of sensors and so many ways sensors can be classified. The type and classification give you some insight into how the sensor works, what it measures, how it measures, and how it can be used. One way sensors can be classified is by the input signal that it measures and by the output signals it produces. *Analog* and *digital* are the most basic sensor classifications. An analog sensor creates a signal that has a continuous value. That value is proportional to the quantity being measured. When measuring physical quantities such as temperature, pressure, and speed, which are all analog quantities, the signals are naturally continuous. A sound sensor is an analog sensor, and the electrical signal it produces is analog. An analog signal can have a range of 0 to 5V DC. Thus the detection of no sound would have an analog signal of 0V, and the maximum sound detected would have a maximum of 5V. Analog sensors will detect a signal of any value within a range, so there can be voltage values between 0 and 5, as illustrated in Figure 3-2. This figure shows the analog sensor signal as a continuous signal between 0 and 5V. Analog sensors will continuously detect changes in the signal.

In order for the microprocessor to do such things as compute values and perform comparisons on the returned signal, the signal is converted to a digital signal by an *analog-to-digital* (A/D) *converter*. Converting the analog signal to discrete values is called *quantization*, as shown in Figure 3-2. The A/D converter will divide the range into discrete values. The maximum number of volts is divided by the possible permutations of the converter determined by the n bits of the converter. An NXT A/D conversion is done by the AVR coprocessor, which uses a 10-bit converter. This means that there are 10 bits that will represent each signal in the sample. The signal has a 10-bit resolution in this case. There are $2^{10} = 1,024$ levels for the A/D converter for this sound sensor. To determine the voltage resolution, divide the overall voltage measurement by the number of discrete values. Here is the calculation:

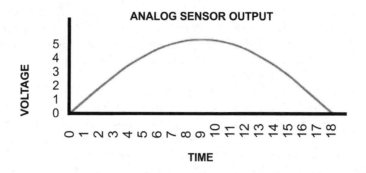

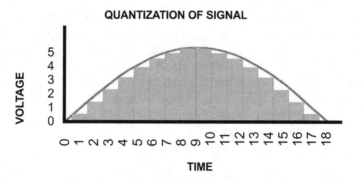

FIGURE 3-2 The continuous signal of an analog sensor and its quantization (digital signal) between 0 and 5 V.

$$5V/1{,}024 = 0.00488V = 4.8 \text{ mV}$$

The voltage resolution is 4.8 mV (millivolts). The greater the number of bits, the better is the digitized representation of the original signal and the lower is the quantization error.

The analog sensor is designed to convert real-world information that isn't electrical into a format that is recognizable by a computer or embedded system. Some sensors do not turn their input directly into digital signals but instead create an analog signal that is converted later. Sensors often have a very low amplitude and therefore require some amplification. The amplification *increases* the value of the signal to better use the full range of the conversion.

Digital sensors produce a discrete output signal that is a digital representation of the quantity being measured. This means that a digital signal produces only discrete (noncontinuous) values that may be outputted as a single "bit" as in a simple switch. When the switch is on, the circuit is closed, and electricity flows through the circuit. When the switch is off, the circuit is open, and electricity does not flow. A switch is an example of a serial transmission that outputs a binary signal that has only two discrete states of "1" and "0" ("ON" and "OFF"). Bits also can be combined to produce a single-byte output of n bits as a parallel transmission. An example of a digital sensor is the built-in optical incremental encoder. It outputs a relative position of the motor to the last position. An encoder coupled with the NXT and a tachometer measures the velocity and direction of the motor. An optical encoder detects movement of the servo motor. We will discuss optical incremental encoders and servo motors in Chapter 4 in our discussion of actuators.

A digital sensor also has an embedded microcontroller that directly processes the signal inside the sensor. The data transmission is also digital, which means that it is not sensitive to cable length, resistance, or impedance and is not influenced by electromagnetic noise. A digital sensor also can return multiple values. Some of these values can be used to manipulate parameters such as calibration values. Analog sensors simply produce the signal output, the voltage reading. That's all. True analog sensors do not have embedded chips, so the A/D conversion is performed outside the sensor.

Analog sensors are more accurate because the original signal is represented with a higher resolution, but these signals can be easily affected by noise or degrade in transmission. Analog signals are also hard to use in calculations and comparisons. However, data loss occurs in converting analog signals to discrete values. Analog values are more accurate, but most of the sensors you will find for your NXT robot are digital or, should we say, analog sensors with A/D converters that are considered to be digital by third-party vendors. Table 3-2 compares some of the attributes of digital and analog sensors.

TABLE 3-2   Comparison of Analog and Digital Sensors

Attributes	Analog Sensor	Digital Sensor
Type of signal	Continuous	Discrete
Accuracy of signal	High accuracy owing to the measure being close to the original signal	Some data loss
Signal conversions to digital	Loses some accuracy when converted to digital	No conversions
Use of signal by microcontroller	Must be converted Difficult to be used in calculations	Ready to be used Ease of use
Signal processing	Processing of signal outside sensor Requires amplification	Outboard microcontroller for processing None required
Signal transmission	Sensitive to degradation and noise	No degradation during transmission
Signal output	Only voltage reading	May contain additional information besides reading

Internal and External Sensors

Sensors measure the factors or characteristics of an environment. That environment can be the internal environment of the robot, measuring its internal state. These sensors are called *proprioceptive*. For a self-balancing robot, gyro sensors are used to measure the angle the robot is tilting and its angular velocity. If the robot is tilting forward, then this number will be positive, and if it is negative, then the robot is tilting backward. The angular velocity is the change in the rotational angle of the gyro. To prevent the robot from leaning forward or backward, the robot's wheels will roll forward or backward. The position of the motor axle and the velocity of the motor are monitored, controlled, and corrected. Detecting 0 acceleration means that the robot is balanced and leveled. The contact sensors (measuring an internal contact between the contact point and the sensor) are used to detect the position of the motor. They help to measure the velocity and acceleration, such as encoders and tachometers. The compass sensor measures the earth's magnetic field and calculates a magnetic heading that reflects the direction the robot is facing. These are examples of detecting the robot's internal state.

 The gyro, accelerometer, and compass sensors for NXT robots are sensors that are physically external to the NXT brick and are attached to the chassis of the robot. They are still considered *proprioceptive* sensors because the sensor is inward-facing to the robot as opposed to the outward-facing sensors, which are discussed next.

Sensors that measure the external environment as it interacts, intersects, and/ or affects the robot are called *exteroceptive* sensors. Exteroceptive sensors can be contact, proximity, or ranging sensors. These contact sensors are used to measure contact between the robot and some other object in the environment. Proximity sensors measure the distance to objects near the robot but that are not touching the robot sensor. Ultrasonic and optical sensors are examples of proximity sensors that use ultrasonic waves and light to measure the distance an object is from the robot. The robot is the point of reference for these types of robot sensors. Another type of sensor that measures physical quantities in the environment, such as surface temperature, pH levels in liquids, turbidity, air pressure, and magnetic fields, are called *environmental* sensors. These quantities do not require the perspective of the robot. Figure 3-3 illustrates the different perspectives of the robot as it relates to proprioceptive, exteroceptive, and environmental sensors.

Active and Passive Sensors

Whereas proprioceptive and exteroceptive sensors are characterize by what is measured, *active* and *passive* sensors describe how the sensors make measurements and how they meet their power requirements. Passive sensors do not require an external source of power because they convert the energy of the external stimulus into the output signal. They directly produce an electrical signal from the input they are measuring. Consider a passive infrared sensor (PIR) used as a motion detector as an example. A PIR sensor measures the infrared light that radiates off objects in its

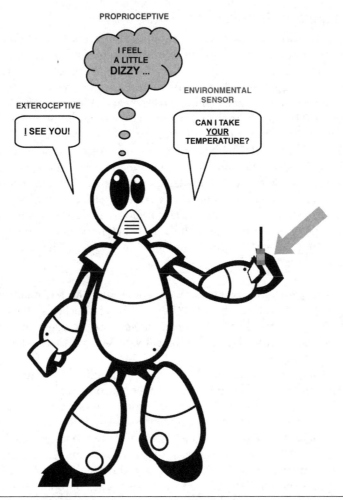

FIGURE 3-3   A robot's perspective in terms of proprioceptive, exteroceptive, and environmental sensors.

field of view. The sensor detects when there is a change in the normal radiation. For example, in Figure 3-4a, a robot enters the field of view of the sensor. The sensor is triggered because the robot breaks the continuous field. What the sensor detects is the disturbance, a "tremor in the force." The sensor is passive because it does not emit an infrared beam. It accepts the incoming radiation without any type of intervention. Other examples of passive sensors are touch, light, and some temperature sensors. Most passive sensors are considered *direct* sensors.

Active sensors, on the other hand, do require an external source of power to produce an output signal. The sensor generates an electrical signal that changes the excitation signal. By passing an electric current or pulse through the excitation signal, the sensor measures the changes in the current that is reflected back. Going back to

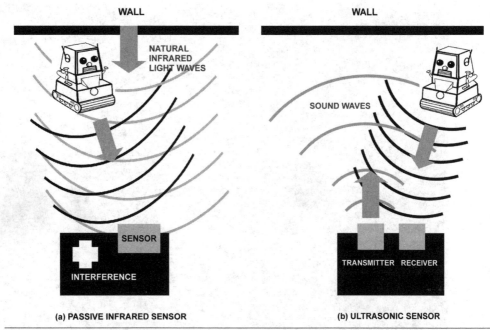

FIGURE 3-4   The difference between passive and active object detection: (*a*) passive infrared sensor; (*b*) ultrasonic sensor.

the motion detector, motion also can be detected by using an active sensor such as an ultrasonic sensor. An ultrasonic sensor generates sounds waves in the ultrasonic frequency range, typically 30 to 50 kHz. The sensor emits a cone-shaped sound waves in the 40-kHz frequency. This frequency is inaudible to people (who hear 20 Hz to 20 kHz) and cannot pass through most objects. The sensor then listens to the sound that is reflected back off the objects in its field of view. The time it takes to transmit and then receive the reflected wave determines the distance to the object. Thus, for a robot detection system, the ultrasonic sensor is continuously sending ultrasonic waves. If something is within the ultrasonic sensor's field of view, the sound waves will be reflected off the new object, returning a different reading, as shown in Figure 3-4*b*.

In some cases there are sensors that have an active and passive mode or active and passive versions. A digital camera is an example of a device that has both an active and a passive mode. A digital camera has an image sensor that converts optical images to electronic signals. In passive mode, the sensor uses the existing or ambient light in its field of view, as depicted in the image in Figure 3-5*a*. The image sensor records or captures the radiation provided. If there is not enough light, the user can select a flash, or the camera detects that there is not enough light and switches to an active mode, and the camera provides a flash. The flash is its own energy source that illuminates the field. It then records the radiation that is reflected from the field. Figure 3-5*b* shows the same image with a flash. This is precisely the meaning of active and passive modes for light sensors. The passive mode measures the ambient light

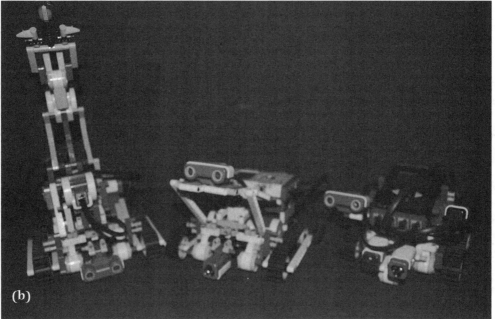

FIGURE 3-5   Demonstration of active and passive light sensors using a camera: (*a*) image taken without a flash; (*b*) image taken with a flash.

rather than the reflected light from the light-emitting diode (LED). In active mode, the sensor will emit its own light source, the LED, and measure the light that is reflected. There is an active version of the infrared sensor that works in a manner similar to the ultrasonic sensor. It is the light version of a sound sensor. An active infrared sensor uses invisible light as opposed to high-frequency sound to scan an area. The light is reflected off the objects in the scan zone and detected by the receiver. Table 3-3 lists these types of sensors and a brief description of each.

TABLE 3-3   Attributes of Internal, External, Active, and Passive Sensors

Category	Sensor Type	Description
What is measured	Internal (proprioceptive)	Sensors that measure the different factors or characteristics of the internal state of the robot
	External (exteroceptive)	Sensors that measure the different factors or characteristics of the external environment as it interacts with the robot
How it is measured	Active (ultrasonic, encoders, RFID)	Sensors that require an external power source because they generate an electrical signal that modifies the excitation signal and then measures the change of the current that is reflected back
	Passive (compass, sound, pH)	Sensors that do not require an external source of power because they convert the energy of the external stimulus into the output signal

There are quite a number of sensors that can be described as both active and passive besides the ones just mentioned. Active and passive sensors each have their advantages and disadvantages, as do all sensor types. Robot team members equipped with different types of sensors are better able to perform and complete tasks more efficiently and under various conditions. For example, in Figure 3-6, a robot team performs motion detection. R1 is equipped with an active infrared sensor, and R2 is equipped with a passive infrared sensor. Passive infrared sensors are more accurate in narrow areas than active infrared sensors but can produce false-positive results when the temperature is high. Active infrared sensors perform better on objects at distances and in low-light environments. Using robot team members with both types of sensors monitoring the same environment under varying climate and lighting conditions will increase the accuracy of a motion-detection system.

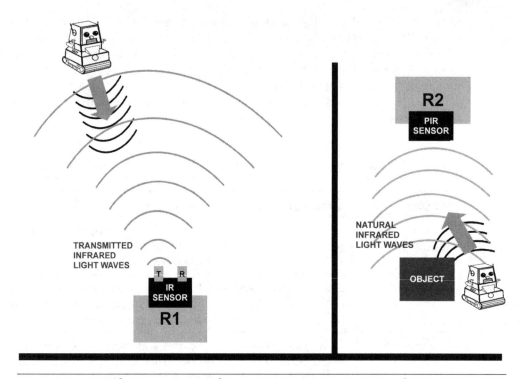

FIGURE 3-6 A robot team motion-detection system using passive and active sensors.

Classifying MINDSTORMS NXT Sensors

Table 3-4 lists the standard LEGO MINDSTORMS and others sensors, their manufacturers, and their classification. Analog, digital, proprioceptive, and exteroceptive, as well as passive and active, describe different characteristics of sensors and are not mutually exclusive groupings. Thus sensors can be described as analog-proprioceptive-passive or digital-exteroceptive-active. And some sensors can be active and passive or at least have an active and passive mode. The standard manufactured LEGO MINDSTORMS sensors are for

- Light
- Touch
- Color
- Sound
- Temperature
- Ultrasonic
- Rotation

MINDSTORMS sensors are both analog and digital. The analog sensors are light, touch, color, sound, and temperature. The digital sensors are the ultrasonic and

TABLE 3-4 Classification of Sensors

Function	Description	Sensor	Manufacturer	A	D	EC	PC
Proximity and presence	Measures the distance the robot sensor is from an object using sound waves, light, or infrared radiation; the distance is determined by the time it takes for the instrument of measurement (wave, light, etc.) to reach the sensor in order for the sensor to take a reading; sensor can be contact or noncontact type.	**Noncontact**					
		Ultrasonic	LEGO	✓		✓	
		EOPD	HiTechnic	✓		✓	
		Magnetic	HiTechnic		✓	✓	
		Optical/magnetic encoders	LEGO/HiTechnic	✓			✓
		Contact					
		Touch	LEGO			✓	
Location	Detects the location of the robot in a fixed point.	GPS	Dexter	✓	✓	✓	
Heading	Measures the orientation of the robot in relation to a fixed point.	Compass	HiTechnic, Dexter		✓	✓	
Acoustic/sound	Measures sound levels.	Sound	LEGO		✓	✓	
Imaging, light, infrared	Collects data from the surfaces of objects.	Color	LEGO/HiTechnic	✓	✓	✓	
		Infrared	HiTechnic	✓	✓	✓	
		Thermal infrared	Dexter		✓	✓	
		Camera	Several	✓	✓	✓	
Angle, motion, acceleration	Monitors changes in speed, angle, and motion.	Gyroscope	HiTechnic		✓		✓
		Acceleration/tilt	HiTechnic		✓	✓	✓
		Angle	HiTechnic		✓		✓
Temperature, pressure	Detects temperature and pressure.	Pressure	Dexter		✓	✓	
		Temperature	LEGO		✓	✓	
Identification	Used to identify objects.	Radio frequency ID (RFID)	Codatex	✓	✓	✓	

Note: A = analog; D = digital; EC = exteroceptive; PC = proprioceptive.

rotation sensors. What seems unusual about how these sensors are classified is that the touch sensor seems to be an anomaly. An NXT touch sensor is either touched or not. When examining the raw values produced by the sensor, the not-touched (close-to-ground) value is close to 180, and the touched (close-to-supply) value is close to 1,023. The nature of the input signal is not the only consideration when classifying sensors as analog or digital. The digital sensor has an embedded microcontroller that processes the signal. The touch sensor does not have an embedded electronic chip. Its signal is not sent directly to ARM7 but to the AVR coprocessor that performs A/D conversion. This is also true for other analog sensors. Once processed, the signal is passed to ARM7 via the I^2C bus interface. The ultrasonic sensor is digital and communicates directly to ARM7 via the I^2C bus. Figure 3-7 shows the hardware block diagram of the NXT brick from Chapter 1. Here we are focused on the input circuits, the Atmel AVR coprocessor that performs the A/D conversion, the I^2C bus interface, and the ARM7 NXT microcontroller. The rotation (optical incremental encoder) sensor is also classified as digital. We will discuss this sensor in the section covering actuators and motors.

The light and rotation sensors are active. The NXT brick has a current generator that controls the power delivery to the active sensors. They supply power for 3 ms, and then the analog value is sampled during the 0.1 ms that comes after. The timing diagram is shown in Figure 3-8. The passive sensors (i.e., touch, light, sound, and temperature) do not require special power. These sensors are also sampled every 3 ms

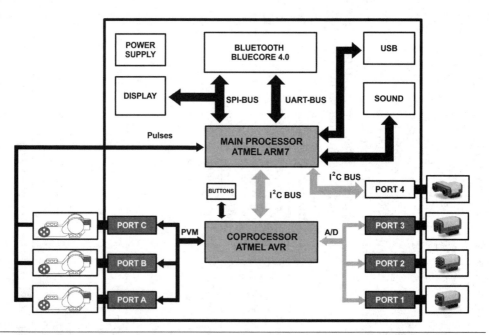

FIGURE 3-7 Hardware block diagram of the input circuit, Atmel AVR coprocessor, I^2C bus interface, and the ARM7 NXT microcontroller.

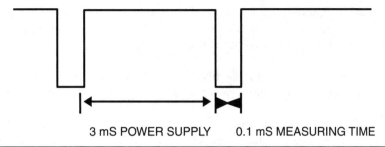

3 mS POWER SUPPLY   0.1 mS MEASURING TIME

FIGURE 3-8   The timing diagram for an analog sensor.

using the A/D converter. All LEGO MINDSTORMS NXT sensors are exteroceptive sensors except for the rotation sensor.

NXT Sensors and the I^2C Bus Interface

The I^2C Digital Protocol is a communication protocol used to connect low-speed devices to computers (in our case, the NXT microcontroller). It is a digital interface that allows external devices (in this case, analog and digital sensors) to send the results of the functionalities they have performed internally to the NXT brick and receive messages from the NXT. The sensors are used as external memory areas. These areas are divided up to easily read and write data that can potentially have device-specific multiple parameters. These parameters can hold information such as device names, calibration values, and startup time. The memory areas have an addressing scheme specifically for each digital sensor that includes:

- The I^2C unique address
- A set of internal addresses to write and read data from

The beginning address and the communication direction bit (with the significant bit first) are transferred.

The NXT microcontroller has four input ports for the sensors, and each port uses the I^2C Digital Protocol running at 9,600 bits/s. Each communication channel has a 16-byte input and output buffer, so a maximum of 16 bytes can be sent or received during each data communication cycle. Each port is enabled to have both digital and analog interfaces. This is the reason why any digital or analog sensor can be connected to any input port. All ports are the same except for port 4. It has a little something extra. Port 4 has the potential to handle high-speed communication (at 921.6 kb/s) as well as bidirectional data communication. This port is for sensors that require a higher communication speed. leJOS NXT has implemented a high-speed I^2C protocol that returns data at a speed of around 2 ms. Keep in mind that response time of the actual sensor factors in the time it takes to initiate a reading to its return to the NXT microcontroller.

The I^2C communication protocol has a master-slave messaging protocol. The sensors are slaves, and the NXT microcontroller is the master that always initiates the messaging. It can send two types of messages:

- A `Read` message that reads one or more bytes from the sensor
- A `Write` message that writes one or more bytes of data to the sensor

Each type of message begins with a header consisting of the sensor address and a register index. The register index is the sensor's internal address, where data are read and written to.

Sensors in the NXT World

In the last section we discussed the different types of sensors that can be used by our robot teams. In this section we will discuss the sensors used in the NXT world. Table 3-4 lists some of the different sensors. It is important to keep in mind what sensors your robots have when designing your robot teams. The capabilities of the sensors used by the robots will determine what your team can and cannot do and under what conditions they perform best and maybe not so well.

We have discussed sensors based on their type (digital or analog), how they take measurements (actively or passively), and the environment in which they make their measurements (internal or external). In Table 3-4 we have also listed the sensors based on their functionality:

- Proximity
- Location
- Heading
- Acoustic and sound
- Imaging, light, and infrared
- Position, angle, motion
- Temperature and pressure
- Identification

Some Are Strong, Some Are Mobile, Some Are Smart

Understanding how their sensors work is very important in determining which robots will be on the team. The confluence of sensors, motors, and end effectors defines the capabilities of the Bluetooth Robotic-Oriented Network (BRON). When determining the sensors or suite of sensors on a single robot, you must consider what sensors work well together, are complementary, and help to accomplish the desired task. One of your main considerations when choosing and placing sensors is to make sure that the sensors do not impede each other's performance or crowd the real estate of the robot.

For example, the compass sensor measures magnetic fields, so the sensor can be affected by a magnet or metal that creates its own magnetic field. The sensor should be isolated from the motors and even the NXT microcontroller. The sensor should be at least 15 cm away from the motor and 10 cm away from the NXT brick.

In expanding the capabilities of a robot by adding sensors and motors, there is a physical limit to the number of sensors that can be connected to the NXT brick. Using a sensor multiplexor (SMUX) will allow multiple sensors to share an input port. MindSensors and HiTechnic both produce multiplexers. The MindSensors multiplexer or port splitter can split a single port into three ports, and the HiTechnic multiplexer splits a single port into four ports. It also supplies a 9-V battery box in order to provide adequate power to each of the attached sensors, which otherwise would be a drain on the battery of the NXT. Again, where will these extra sensors go? A suite of sensors that spans multiple robots requires a different approach when making evaluations. Where two sensors may interfere with each other's performance or may not fit owing to the limited real estate or lack of additional input ports on a single robot, locating one of the sensors on another robot on the team may be ideal. What you are trying to achieve is the right combination of sensors that will expand capabilities and make up for the limitations of other sensors. A suite of sensors can contribute to the team's behavior, making the team more autonomous.

What the Sensors Can Do and Cannot Do

The attributes or characteristics of the sensors describe the reading ranges, how long it takes the sensor to respond to a stimulus, its overall accuracy, and so on. What is its resolution and repeatability? With the microcontroller connected to several sensors, the voltage level and power consumption must be considered. This may determine how long an individual sensor is used or how long other sensors with a high power consumption are used. An awareness of these attributes will help you to see the limitations of sensors and which sensors can be used to compensate for those limitations. You also can use these attributes to compare the quality of sensors. There are several manufacturers of sensors, and there are a few sensors that are manufactured by more than one company (e.g., color by HiTechnic and LEGO). Table 3-5 presents a comparison of compass sensors. You can see that resolution could depend on the version of the firmware in use. Comparing these characteristics will help you to decide which sensor will work best for your robot team.

Characteristics of Sensors

Sensors have a number of characteristics:

- Resolution
- Range
- Linearity
- Accuracy
- Response time

TABLE 3-5 Comparison of Compass Sensors

Compass Manufacturer	Resolution	Refresh Rate	Range	Multiple Readings	Housed
HiTechnic	1 degree	100 × per second	0–359	N/A	LEGO standard
MindSensors	0.01 degree NBC, Robot C 1.44 degrees NXT-G	N/A	0–359	Byte, int, float	Nonstandard

- Sensitivity
- Frequency response
- Reliability
- Repeatability
- Dimension/weight

and these characteristics are listed in Table 3-6. The sensor characteristics of a specific sensor are supplied by the sensor manufacturer. Not all manufacturers that supply technical information will necessarily supply this list. They may supply other information that they feel is important to their customers. Some of these traits may seem important to know right off, such as accuracy, response time, range, and resolution. Others may not be so important, such as linearity.

Range and resolution are two of the most common characteristics that many users are at least at first most interested in. Range is the difference between the smallest and the largest outputs that a sensor can produce or the inputs in which the sensor can operate properly. These values can be absolute or a percent of the appropriate measurement. For ultrasonic sensors, the output range is 0 to 255 cm because distance is being measured, whereas the output range for a compass is 0 to 360 degrees. A light sensor returns a value between 0 and 1,023 to the microcontroller. In a well-lit room, the darkest reading may be 478, and the brightest reading may be 891. In a dimly lit room, the darkest reading may be 194, and the brightest reading may be 445. Thus a percentage scale is used in which 0 is dark and 100 is bright. In the well-lit room, the reading will be 47 to 87 percent, and the dimly lit room will have a reading of 19 to 43 percent. But the sensor can be calibrated to the light of that particular environment, making the darkest reading (at 19 percent) the 0 percent reading and the brightest (at 87 percent) the 100 percent reading. Resolution is the minimum step size within the measurement range of the sensor. The resolution of the LEGO MINDSTORMS Ultrasonic sensor is 1 cm.

What is linearity? It is the relationship between the input and output variations in analog sensor readings. Linearity can be used to predict the readings of a sensor (based on input) and to determine the accuracy and measurement error. If a sensor's output is linear, then any change in the input at any point within its range will produce the same change in the output. The output is proportional to input over its entire range.

TABLE 3-6   The Characteristics of the Basic LEGO Sensors Used in the BRONs in This Book

Characteristic	Description
Resolution	The smallest change of input that can be detected in the output; it can be expressed as a proportion of the reading or in absolute terms.
Range	The maximum and minimum values that can be measured.
Linearity	The extent to which the actual measurement of a sensor departs from the ideal measurement.
Accuracy	The maximum difference that will exist between the actual value and the indicated value of the output of the sensor; it can be expressed either as a percentage or in absolute terms. (A is the actual value and E is the expected value.) $$1 - \frac{A - E}{E}$$
Response time	Time required for change from its previous state to a final settled value.
Refresh rate	How often the sensor takes a reading.
Sensitivity	Change required to produce a standardized output change.
Reliability	Repeatability and consistency of the sensor.
Repeatability	The ability of a sensor to repeat a measurement when put back in the same environment.
Dimension/weight	The size and weight of the sensor.

The graph of the slope of output versus input would be a straight line. For example, if the ratio of input to output is 1:1, then if there is an increase in the input (the stimulus) of the sensor by 2, that would be reflected in the output. Ideally, sensors are designed to be linear, but not all sensors are linear when it comes to actual values. Figure 3-9*a* shows the ideal linear relationship between input and output and the *measured curve* of a fictitious sensor, as well as where the maximum error occurs. Figure 3-9*b* shows the linearity of the LEGO MINDSTORMS Ultrasonic sensor, which is pretty linear. When measuring the distant to an object 2 to 64 cm from the sensor using 63 samples, the average error was 1.079 cm. Often nonlinearity is specified by a percentage:

$$\text{Nonlinearity } (\%) = (D_{in(max)}/I_{f.s.}) \times 100$$

where $D_{in(max)}$ is the maximum input deviation, and $I_{f.s.}$ is the maximum full-scale (f.s.) input. But the linearity of a sensor, as with the ultrasonic sensor, will depend on the conditions in which the readings were taken. Under the best conditions, a sensor may have very linear readings. In an environment hostile to the performance of the sensor (if there were multiple ultrasonic sensors transmitting sound waves at the same time),

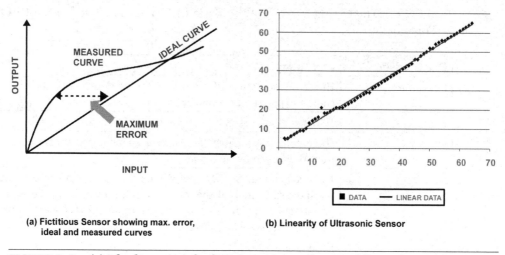

(a) Fictitious Sensor showing max. error, ideal and measured curves

(b) Linearity of Ultrasonic Sensor

FIGURE 3-9 (*a*) The linearity of a fictitious sensor showing where the maximum error occurs. (*b*) The linearity of the LEGO ultrasonic sensor.

the performance of the sensor would degrade, and readings would not be linear. In addition, the material of the object and the degree of the angle the object is to the sensor also may affect readings.

The response time of a sensor may not be totally up to the sensor. It is the time that is needed to observe the change in the output as a result of the change in the input. Depending on how the reading is used, it may require a very fast response time by the robot. The robot taking the reading may have to send the output to another robot on the team before any action is taken, and that action may be time-critical. Thus we must include the response times of the sensor and the Bluetooth device. The ultrasonic sensor's response time is based on how long it takes for the sound waves to travel (34.3 cm/ms at 20°C) to the object and reflect back to the receiver. That time is added to the time it takes for the I^2C bus to return the value to the microcontroller (9,600 bits/s) and the time it takes for the Bluetooth device (also response time) to send this signal to the waiting robot.

Accuracy, Limitations, and the MINDSTORMS Ultrasonic Sensor The ultrasonic sensor is a perfect sensor on which to base a discussion of accuracy and limitations. There are a number of ways to address accuracy as far as range is concerned, and these are the basis of some of the sensor's limitations. Other limitations have to do with the material of the object and the angle of the object in reference to the sensor, as we mentioned earlier. The accuracy of the sensor is determined by how close the actual value is to the expected value. This calculation is shown in Table 3-6, where A is the actual or measured value and E is the expected or true value. Thus an object located 60 cm from the ultrasonic sensor should produce a reading of 60 cm. The accuracy of the ultrasonic sensor is documented at ±3 cm in a range of 0 to 255 cm.

The sensor shows more accurate performance within certain ranges. As indicated earlier when we described the ultrasonic sensor's linearity, the sensor cannot make a reading for objects at a distance under 3 or 4 cm because of the time it takes a sound wave to travel and return. Objects measured at distance above 180 cm will return a weak signal, which also may be unreliable. What about objects on either side of the sensor? Can they be detected? This is called *directivity*, measurement of the directional characteristic of a sound source, which in this case is a 30-degree cone. Figure 3-10 shows the range limitations of this sensor. There are objects outside the

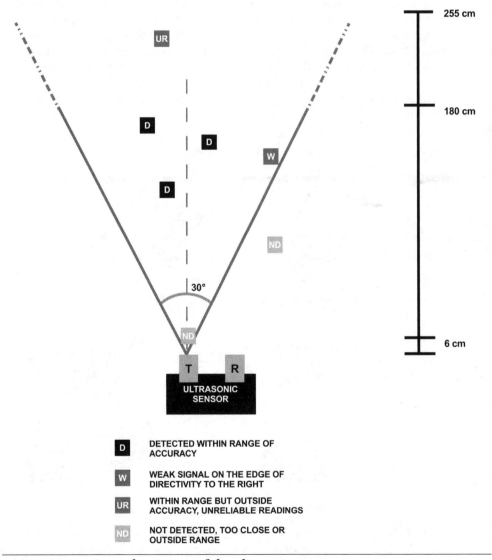

FIGURE 3-10   Range limitations of the ultrasonic sensor.

180-cm range and its directivity. Objects that are too close may not be detected, or the readings may be inaccurate.

Thus we can see how the distance an object is from a sensor can affect the sensor's accuracy. What about the angle of the object in reference to the sensor? How will that affect accuracy? Figure 3-10 also shows that an ultrasonic sensor mechanism has a transmitter and a receiver. The transmitter transmits a 40-kHz sound wave. The sensor performs best when the sensor is high on the robot, avoiding the sound waves hitting the floor and hitting perpendicular to the object. Figure 3-11 shows the various limitations of using sound waves when detecting objects based on reflection or absorption of sound waves. Figure 3-11a shows two scenarios of ultrasonic sensors sending out sound waves (or pings) that are reflected off an object (or wall). In the first scenario, the sound wave on the left is returned first, and it is too short. This is called *foreshortening* (a). In the second scenario, the sound wave hits the surface at an acute angle and then bounces away. Odd-shaped objects reflect the sound waves in ways that are not expected (b). *Cross-talk* occurs when multiple ultrasonic sensors are used simultaneously (all using the same frequency), and the signals may be read by the wrong receiver (c).

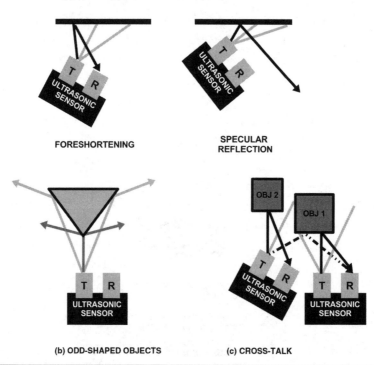

FIGURE 3-11   (a) Foreshortening and specular reflection of sound waves. (b) Odd-shaped objects reflect waves in unexpected ways. (c) Cross-talk causes reading confusion.

With the wide cone and the way sound waves spread out, no distinction can be made between a wide object and a skinny one. Compare this between the LEGO ultrasonic sensor and the MindSensors active infrared long-range sensor, which has a range of 20 to 50 cm. The MindSensors sensor also has an emitter and a receiver, but it uses a single beam of infrared (IR) light and triangulation to detect the distances of objects. This is shown in Figure 3-12. With a narrow beam, an object can be located if the sensor is directly in its field of view with a directivity of only 5 degrees. This narrow beam can be used to detect doorways, whereas the ultrasonic sensor may detect the door frame but not the opening. The MindSensors sensor also can be used to detect the widths of objects. A team equipped with both types of proximity sensors could better navigate a path through doorways and around a room with variously sized obstacles (Figure 3-12). Keep in mind that the ultrasonic sensor can detect up to eight objects (returning multiple readings), but not all software provides this option.

Special Sensors Give That Extra Something

Third-party vendors have developed some of the most interesting sensors to add something really special to any team of robots. These sensors run the gamut of types from color and proximity to temperature and pressure. Some third-party vendors

TEAMWORK USED FOR NAVIGATION

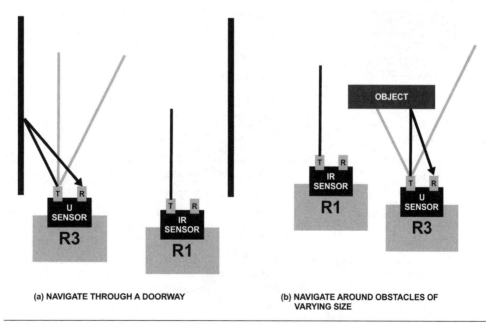

(a) NAVIGATE THROUGH A DOORWAY   (b) NAVIGATE AROUND OBSTACLES OF VARYING SIZE

FIGURE 3-12   Single-beam infrared sensor using triangulation working with an ultrasonic sensor.

such as HiTechnic and CODATEX have been declared official third-party vendors, developing endorsed hardware accessories and software support. Table 3-7 is a list of vendors that have developed NXT-compatible hardware accessories including sensors.

TABLE 3-7   Third-Party Vendors That Manufacture NXT-Compatible Hardware

Third-Party Vendor	Description	Official Vendor
HiTechnic	Variety of sensors Multiplexers Sensor kits IR electronic ball Connector cables	Yes
Hitec	Servos used by LEGO and Tetrix Robotics Systems	No
MindSensors	Variety of sensors Real-time clock Servos HID Multiplexers Controllers Sensor kits Port splitters Ball casters Cables Plugs/sockets	No
CODATEX	Radiofrequency ID (RFID) sensors and transponders	Yes
Vernier	Official LEGO NXT adapters Variety of scientific sensors	No
Dexter Industries	Variety of sensors Solar panels WiFi sensors	No
Techno-Stuff Robotics	A few sensors designed for RCX but NXT-compatible with cable Relays Holonomic wheels	No

The RFID sensor system, compass sensor, and pH sensor probe are three third-party sensors that we will be using in one of the BRONs we develop in this book.

Third-Party Sensors Used in Our CSI BRON

There may be over 100 different third-party sensors available to developers to enhance the capabilities of a team of robots. We used three third-party sensors for our CSI BRON:

- The RFID sensor and transponder system
- The compass sensor
- The pH sensor probe

The radio frequency identification (RFID) sensor and transponders system for the LEGO MINDSTORMS NXT is manufactured by CODATEX. RFID sensors are used to identify an object and store information about it when the object is not in the line of sight of the sensor. An RFID system has a reader, a transponder or tag, an antenna, and a host interface that is connected to the reader, as shown in Figure 3-13. The RFID sensor transmits a low-frequency radio signal (125 kHz) through the integrated ferrite antenna. The electromagnetic wave activates the transponder by inducing energy in the coil of the transponder. Then the transponder, which is passive, starts to modulate the signal according to its number (the transponder's ID number), which is stored in nonvolatile memory. Modulation is the process of mapping a wave to contain the information that is to be transmitted, thereby converting the wave to a signal. The modulated signal can be detected (demodulated and decoded) by the RFID sensor.

The ID of the transponder is a 5-byte number. There are 2^{40} possible combinations for IDs. In addition, the reader has a number of reading modes:

- Single
- Continuous
- Stop

FIGURE 3-13   CODATEX RFID sensor and transponder system.

With a single read, if the transponder (tag) is close enough to the antenna, the data will be read. There is a maximum read rate of 3 reads/s. Once the single read occurs, the sensor goes into sleep mode. A continuous read means that readings are performed continuously on available transponders. In this mode, as many as 10 reads can occur per second. If there is a 2-second delay (no readings occur), then the sensor goes into sleep mode. Stop mode causes the sensor to stop reading and go into sleep mode.

The sensor can detect the signal regardless of the presence of objects (nonmetallic) between the sensor and the transponder. But the object has to be within 3 cm of the sensor depending on the size of the transponder. Other limitations can cause the sensor to detect a bad, low, or no reading. Certain orientations of the transponder's coil will cause a low signal or no signal detection at all. Detection also can be disturbed by electrical noise from motors, electrical switches, and fluorescent lights. The farther away the transponder is from the sensor, the more electrical noise will be a factor in taking a reading. However, the reading is refreshed or updated 100 times a second.

HiTechnic's digital magnetic compass sensor measures earth's magnetic (geomagnetic) field and calculates a heading angle. A magnetic compass works by detecting and orienting itself in the very weak magnetic field on the surface of the earth thought to result from rotational forces of the liquid iron in the earth's core. The sensor returns a value from 0 to 359 that represents the current heading angle of the robot, as illustrated in Figure 3-14.

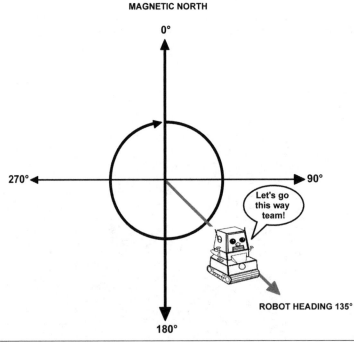

FIGURE 3-14 The angle heading of a magnetic compass.

The sensor must be mounted horizontally on the robot. The compass also should be away from anything that emits electromagnetic fields, such as motors, transformers, inductors, and so on. Also, large conductive items can significantly alter magnetic fields (e.g., cars, refrigerators, etc.). Either will affect the reading. To minimize interference, the sensor should be placed at least 6 inches (15 cm or 19 studs) away from motors and at least 4 inches (10 cm or 13 studs) away from the NXT brick.

The Vernier pH sensor measures the acidity of an aqueous (liquid) substance. Acidity is measured on a scale of 0 to 14 pH units, with highly acidic at one extreme of the scale (pH 0) and basic at the other extreme of the scale (pH 14). Table 3-8 lists the

TABLE 3-8   The pH Scale and Representative Substances

ph Value	Example Solution
14	Liquid drain cleaner
	Caustic soda
13	Bleaches
	Oven cleaner
12	Soapy water
11	Household ammonia (11.9)
10	Milk of magnesium (10.5)
9	Toothpaste (9.9)
8	Baking soda (8.4)
	Seawater
	Eggs
7	Pure water
6	Urine (6)
	Milk (6.8)
5	Acid rain (5.6)
	Black coffee (5)
4	Tomato juice (4.1)
3	Grapefruit and orange juice
	Soft drinks
2	Lemon juice (2.3)
	Vinegar (2.9)
1	Hydrochloric acid secreted from stomach lining (1)
0	Battery acid

pH scale and examples of substances with each pH value. A substance that is neither acidic nor basic is neutral (pH 7). Thus substances with a pH value < 7 are acids, and substances with pH values > 7 are bases (alkaline). Each whole pH value below 7 is 10 times more acidic than the next-higher pH value. So a substance with a pH value of 4 is 10 times more acidic than a substance with a pH value of 5 and 100 times more acidic than a substance with a pH value of 6 (higher pH as value decreases). On the other hand, a substance with a pH value of 9 is 10 times more alkaline than a substance with a pH value of 8. Chemicals that are at the extremes of the scale [highly alkaline (pH 14) and highly acidic (pH 0)] are called *reactive*. They can cause severe burns, such as battery acid (acidic) and household drain cleaners that contain lye (basic). Mixing chemicals with water (pure water is neutral with a pH of 7.0) can result in an acidic or basic solution. For example, vinegar is an acidic substance, whereas laundry detergent is basic.

The manual that accompanies a sensor or probe may list these as possible test experiments:

- Household acids and bases
- Acid-base titrations
- Monitoring pH change during chemical reactions or in an aquarium as a result of photosynthesis
- Investigations of acid rain and buffering
- Investigations of water quality in streams and lakes

We just thought the pH sensor was a cool sensor and decided to use it for one of our BRONs.

This probe and all the sensors and probes manufactured by Vernier were not made for the LEGO MINDSTORMS NXT. They are used manually to perform scientific experiments. An NXT adapter was developed to allow Vernier sensors and probes to be used with the LEGO MINDSTORMS NXT robotic system. The sensor is plugged into the adapter, and the adapter is then plugged into any input port.

leJOS (Java) Support for Third-Party Sensors

leJOS (Java) has defined a number of classes to be used to program sensors, including third-party sensors. Table 3-9 lists the classes and methods for the HiTechnic compass and RFID sensors.

leJOS does not define classes for all sensors for NXT. For a sensor without a class, `ClassSensorPort` can be used to create an instance of a generic sensor. Using this approach, the raw value of the sensor is accessed. Call these methods to set the type and mode of the sensor:

- `void setMode(int mode)`
- `void setType(int type)`
- `void setTypeAndMode(int type, int mode)`

TABLE 3-9   leJOS (Java) Classes That Support the Compass and RFID Sensors

CompassHTSensor Method	Description
getDegrees()	Returns the directional heading in degrees
getDegreesCartesian()	Cartesian coordinate systems increase from 0 to 360 counterclockwise, but compass readings increase clockwise
resetCartesianZero()	Changes the current direction the compass is facing into the zero angle for the method getDegreesCartesian()
startCalibration()	Starts the calibration for the compass
stopCalibration()	Ends calibration sequence
RFIDSensor Methods	**Description**
protected java.lang. String fetchString (byte register, int len)	Ensure the device is awake before talked to
byte[]getSerialNo()	Get the serial number of the RFID sensor
int getStatus ()	Read the status of the sensor
byte[] readTransponder (boolean continuous)	Read a transponder ID
long readTransponderLong (boolean continuous)	Read a transponder ID as long
void startBootLoader ()	Enter boot loader mode
int startContinuousRead ()	Start continually reading from the sensor
void startFirmware()	Start the firmware on the RFID sensor
int startSingleRead ()	Start a single read from the sensor
int stop()	Send a stop command to the sensor
void wakeUp()	The sensor will go into a power save mode after a short time.

We used **MODE_RAW** to set the mode of the sensor. Then call:

```
int readRawValue( )
```

which returns the raw value of the sensor. This raw value will have to be converted to the measurement within the range of the sensor. For example, since there is no class for the pH sensor, we used the **SensorPort** to access a sensor reading. For the pH sensor, the voltage of the sensor is multiplied by the slope (–3.838) and then added to the intercept (13.720) to produce the measure. The slope and intercept are supplied by the manufacturer. To calculate the voltage, the raw reading produced by the probe is divided by the possible permutations of the converter determined by the *n* bits of the converter times 5.

$$(Reading/2^{10}) \times 5 = voltage$$

$$Voltage(-3.838) + 13.720 = measurement$$

Measurement will be an integer in the range of 0 to 14.

LabVIEW Support for Third-Party Sensors

LabVIEW has built-in support for HiTechnic and Vernier sensors. Both sensor types can be accessed from the Functions palette. Figure 3-15 shows how to access Vernier sensors from the Functions palette.

Drag the Vernier Read block to the block diagram workspace. The Vernier Read block is a polymorphic VI. It will allow you to select the appropriate sensor from a drop-down list. Click on the down arrow next to 5g to see the list of all the Vernier sensors, as shown in Figure 3-16.

Access HiTechnic sensors from the Functions palette, as shown in Figure 3-17. From HiTechnic Read, all the available HiTechnic sensor blocks are accessed. For other third-party sensors, use an I^2C communication generic sensor VI, as shown in Figure 3-17.

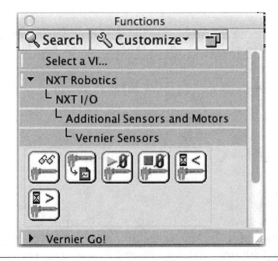

FIGURE 3-15   Accessing Vernier sensors from the Functions palette.

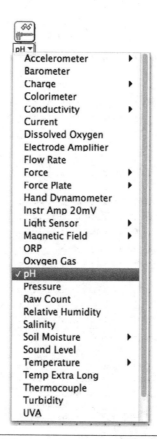

FIGURE 3-16   List of Vernier sensors as accessed from the Vernier Read block.

FIGURE 3-17   LabVIEW's generic sensor VI.

NXT-G Support for Third-Party Sensors

For NXT-G support, the block for third-party sensors is imported. Here are the simple steps:

1. Download the NXT-G block into a directory.
2. Start NXT-G, and select "Block Import and Export Wizard."
3. Click "Browse," and select the directory that contains the subdirectory of the downloaded block to be imported.
4. Highlight the directory, and then click "Import."

The block is then imported to the complete palette in the "Advanced" option, as shown in Figure 3-19. There you see the blocks for the RFID sensor, compass sensor, and Vernier sensor. Figure 3-20 shows the blocks and configuration panels for the RFID, compass, and pH sensor probes.

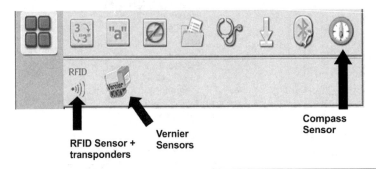

FIGURE 3-19   The complete palette "Advanced" option showing the blocks for third-party vendors.

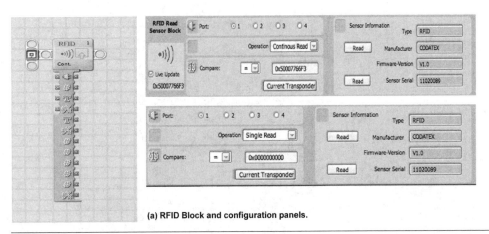

(a) RFID Block and configuration panels.

FIGURE 3-20   Configuration panels for (*a*) RFID.

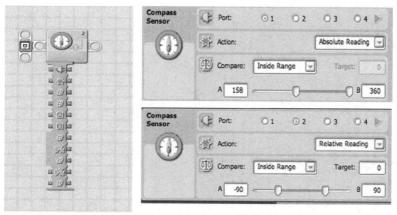

(b) Compass Block and configuration panels.

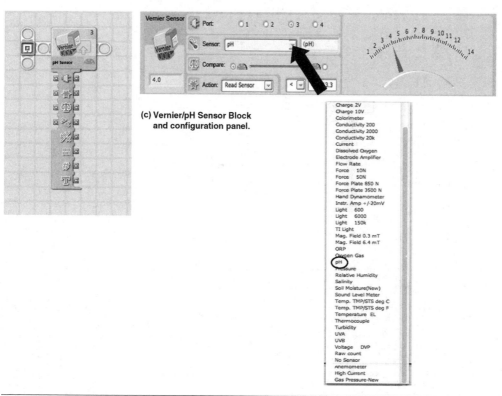

(c) Vernier/pH Sensor Block
     and configuration panel.

FIGURE 3-20   Configuration panels for (*b*) compass, and (*c*) pH sensor probes.

Chapter 4

Creating a Team of Movers and Shakers

The Lost Scrolls of Robotics: #4

You should see me do that when I've got my powerchip.

—Skywarp, *Transformers*

Robots work in an environment. And they can work in an environment because of the interaction they have with that environment. Robots get input from their sensors, and the output is some form of activity in the environment. Without the input from its sensors, a robot does not know the state the environment is in, what is going on, or the current conditions. With this information, a decision is made to do something—some activity takes place. The activity will be implemented by the actuators of the robot team members. One robot will do one part, and another robot will do another part. When assembling a team, you make an assessment of the environment and then examine the capabilities of possible team members, their sensors and their actuators. An actuator is a mechanism that allows an agent, in this case a robot, to act on the environment.

Motors: The Output Transducer

Actuators are a type of transducer—an output transducer. We discussed input transducers in Chapter 3. Transducers convert an electrical signal into another form of energy. The very first figure in Chapter 3 showed the conversion of sound waves by an input transducer to an electrical signal in which the signal could be measured in terms of loudness or softness. Figure 4-1 shows the conversion of an electromagnetic sound wave back to an acoustic wave performed by an output transducer—a loud speaker. Other types of actuators convert electrical signals to electromechanical energy, such as converting a signal to energy that causes a motor to rotate. Examples of such actuators include:

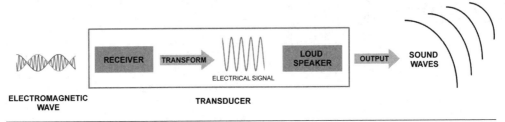

FIGURE 4-1 The conversion of an electromagnetic sound wave to an acoustic wave performed by an output transducer, a loudspeaker.

- Potentiometer when used to measure a position
- Accelerometer
- Linear and rotary motors

In this chapter we will focus on rotary motors. Rotary motors are used as joints for robotic arms and end effectors and to make robots mobile with legs, wheels, and tracks. There are different types of motors; some are more powerful than others, some are good for moving large robots, and others are better at more intricate motor skills. When choosing your robot team members, you have to keep in mind what they will be doing. Do they have motors with the torque necessary for picking up an object and transporting it from one location to another? How much weight can they carry? What type of mobility do the robots require? What terrain will the robots operate on? And what types of motors will the robots need to move around on the terrain?

Indoor and Outdoor Robots

As we said earlier, robots work in an environment. Most robots are programmed to work indoors. Indoor environments are safer and more predictable. If we are in total control of the environment in which the robot team will work, then the challenges are well thought out, and our team will easily be able to perform their task. Even indoor environments can introduce challenges, though. Table 4-1 lists the different types of environments and some of the factors that should be considered.

What is consistently challenging for either environment will be navigation around obstacles and the terrain. We start to address navigation problems by using different types of sensors, such as ultrasonic and infrared sensors. Sensors give us information about the surroundings, such as that an object is there, maybe the size and height of the object, and how far or how close the object is from the robot or from a team member. Once that is done, though, and a course of action is determined, does the robot have the ability to implement it? If the robot is to roll over an object, such as perhaps a small piece of clothing or a doorway threshold in an indoor environment or a curb in an outdoor environment, does the robot have the torque to pull its own weight over it?

TABLE 4-1   Different Types of Robot Environments and Some Important Factors

Robot Environment	Description	
Indoor (most common)	Controlled	Include mazes
		Highly constrained, few surprises
		Robots have a specific task
		Drive trains are simple, using wheels directly connected to motors
	Uncontrolled	Robots roam unconstrained or encounter moderate-constraint terrain issues such as carpet, raised door thresholds, carpet-to-floor and floor-to-carpet transitions, and stairs.
		Ground clutter
		Obstacle avoidance and navigation
Outdoor (complicates robot design)	Weatherizing	Protection from:
		Temperature extremes
		Rain/moisture
		Humidity
	Protection from dirt	Dirt stays away from circuitry and gears
	Vibration damping	Lessen the effect of vibration on mounted motors and circuit boards so that they do not come loose

Dealing with Terrain Challenges

The challenges of terrain can be addressed by steering and driving systems, the type of mobility, and the torque and speed of the robot. Some terrains are better negotiated with legs, especially if climbing is needed (as with stairs). Wheels and tracks are better with other types of terrain. If an object is to be transported, which member of the team will be given that task? The one with legs, wheels, or tractor wheels?

The size and weight of the robot become a factor in making such choices. Smaller robots may have fewer problems in controlled environments. They are fast and can easily get around small obstacles. On the other hand, if a small robot must get to waypoint A from waypoint C in a certain time frame and it is known that a thick carpeted area exists between those two waypoints, the robot's batteries may become depleted while supplying its motors with a high constant current to maintain speed at the cost of torque. The wheels are rotating, but the robot is not going very far. In this situation, one would choose a team member that is larger and has bigger wheels with more powerful motors. Such a robot is less affected by the terrain in this case. However, the bigger, heavier robot also will have some challenges here. Larger robots draw more current. Longer distances mean drawing more current from the batteries, which are already being taxed. When the motor stalls, it will draw a large amount of current. Does the larger robot have indirect drive? This can help with lessening the load on the motors.

What size wheels are needed? Are treads better across any type of terrain? Treads are cool looking. They look like they mean business. They are the obvious choice in outdoor areas with tough terrains, but they can complicate the mechanical design. Large wheels with four-wheel drive are also an obvious choice, but with larger wheels there will be a decrease in the maximum available torque. Legs (not just human bipedal legs but legs that simulate other animals such as insects) are much more complicated mechanical devices for mobile robots and are good in very narrow circumstances. They are good for very tricky, inconsistent terrains, but they require a lot of servos and, more important, complex programming. Do you have any bipeds on the team? Whatever the mobility of your team members, considering the torque and speed capabilities will depend on the motors the robots are using.

Direct-Current Motors vs. Servo Motors

Direct-current (DC) motors and servos give our robots physical action, such as the ability to walk, roll, and pick up and manipulate objects. But what is the difference between a DC motor and a servo motor? Under what circumstances should you pick a team member equipped with servos or a team member that uses predominantly DC motors for mobility? Or under what circumstances should you pick a team member that uses servos or DC motors for the robot arm and gripper? First, you should know the difference between the motor types and their basic capabilities.

Since one of the components of a servo is a DC motor, let's discuss the DC motor first. When you turn on a DC motor, the DC current causes the motor to rotate continuously. What's rotating? Well, the motor has two parts. One is stationary, and the other part rotates. The stationary part is called a *stator*—the motor's "can" or housing. The rotating part is called an *armature* or *rotor*. The way they work will depend on the type of DC motor (i.e., brushed or brushless), but since the servo motor has a brushed motor, let's talk about that one. A rotor continuously rotates around an axle by flipping an electric current. This is accomplished by the brushes and the commutator, a plate around the axle. This flipping in the magnetic field that is produced inside the can causes the rotor to turn. Another current is created as the motor turns faster that has an opposite polarity to the current from the battery. These two opposite currents mean that the motor will draw less current, balancing out and causing the motor to attain a steady state. When the motor has "no load," it is the most efficient, but it is also at its weakest as far as its ability to work. The motor is the strongest when it is working the hardest, and that's when all current flowing is attempting to make the rotor move when it is not moving. This is when there is maximum torque.

Controlling Speed and Torque

The rotation of the motor's shaft supplies speed and torque to the robot. Speed and torque are at odds; more speed means less torque, and more torque means less speed. Speed can be measured as revolutions per minute (rpm). The revolutions per minute for a DC motor can be from thousands to tens of thousands. The speed or velocity (speed and direction) of a given motor can be computed as follows:

$$V_R = (V_M D\pi)/60$$

where V_R is the velocity of the motor, V_M is revolutions per minute of the motor, and D is the diameter of the wheel times π (3.141). For example, a robot that has 4-inch wheels and a DC motor running at 3,000 rpm will have a theoretical maximum speed of 628.2 in/m:

$$628.2 \text{ in/m} = (3,000 \text{ rpm} \times 4 \text{ in} \times 3.141)/60$$

If you increase the diameter of the wheel by an inch, from 4 to 5 inches, you will increase the velocity of the motor from 628.2 to 785.25 in/m, an increase of 157 more inches in a minute.

The speed of the DC motor can be easily controlled by controlling the power level or the voltage supplied to the motor. The higher the voltage, the faster the motor tries to go. A way to control the voltage and therefore the speed of the motor is by using pulse-width modulation (PWM). With PWM, the operating power to the motors is turned on and off to modulate the current that is going to the motor. The ratio of the "on" time to the "off" time is what determines the speed of the motor. Switching the power on and off fast enough makes it seem that the motor is slowing down without stuttering. When using this approach, though, not only is there a reduction in speed, but there is also a proportional decrease in torque.

Torque is the angular force that produces motor rotation measured in pounds per foot, ounces per inch, or newtons per meter. Torque is not a constant value. Torque can take on different values depending on the given information or condition, for example, the stall torque, which is the torque when it is at its maximum. This occurs when the most amount of torque is used to make the motor rotate from a standstill state called the *stall torque*. A *full-load torque* is the amount of torque needed to produce the rated power (horsepower) at the full speed of the motor. The stall torque is often higher than the full-load torque. Sometimes these values are supplied by the manufacturer on the data sheet for the motor. Sometimes other information is supplied, and then you can calculate the stall and full-load torque values.

$$\text{Full-load torque} = (\text{horsepower} \times 5,252)/\text{rpm}$$

Horsepower is measured in watts, and the motor's horsepower rating should be supplied by the manufacturer.

$$\text{Stall torque} = \text{power}_{max}/\text{rpm}$$

No-load speed is at the other extreme; it is a measure of the motor at its highest speed, when it is rotating freely with no torque. These are the two extreme performances of the motor. Between the stall and no load, there is the rated, or nominal, torque, which is the maximum torque that ensures continuous operation without problems. This is approximately half the stall torque. When your robot team members are executing a task, the *startup torque*, the amount of torque required by a robot to perform a task, should be around 20 to 30 percent of its maximum torque. Figure 4-2 shows the relationships among power, torque, and velocity.

If you want your motor to work at its longest running time, you want it to be at its highest speed, not its highest power but still be efficient.

Another way to reduce the speed of the motor without reducing the voltage and therefore the torque is by using gears. With a gearbox, the rotational velocity can be increased or decreased along with a corresponding increase or decrease in torque. Along with this, a number of other things can be accomplished when transmitting power through the use of gears:

- Change the rotational direction
- Change the angle of rotation
- Convert rotational motion into linear motion
- Change the location of rotational motion

The Tetrix DC motor uses a gearbox and is called a *gearhead DC motor*. Figure 4-3 shows this DC motor. The gearbox is on the shaft end of the motor. The gearbox makes the output of the motor or the turning of the shaft slower and more powerful without lowering the voltage.

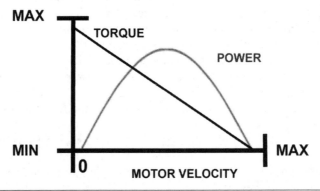

FIGURE 4-2   The relationships among power, torque, and velocity.

DC MOTOR

GEARHEAD

FIGURE 4-3 The DC motor and gearbox on the shaft end.

Gearing Up and Down, Round and Round

Gearing down means using gears to slow down the output of the motor before it is applied to a shaft. How is this done? Well, start with two gears, one small and one large. Consider Figure 4-4a. The small gear is called the *pinion* (gear *X*), and it has 16 teeth. The large gear (gear *Y*) is called the *wheel*, and it has 32 teeth, double gear *X*. The *ratio* of the gears is the number of teeth on the wheel divided by the number of teeth on the pinion. A ratio of 2 means that for each revolution of the pinion, the wheel will make a half turn. Thus, if a motor is connected to the pinion and the output shaft to the wheel, the motor speed is cut in half at the output shaft. On the other hand, the torque output is doubled. The speed and torque of the wheel can be calculated given the speed or torque of the pinion:

$$T_W = e(T_P R)$$

$$S_W = e(S_P / R)$$

where T_W and S_W are the torque and speed of the wheel, respectively, and T_P and S_P are the torque and speed of the pinion, respectively, R is the gear ratio, and e is the gearbox efficiency, which is a constant between 0 and 1. For example, for a pair of gears:

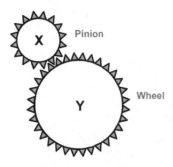

(a) PINION, 16 TEETH GEAR   WHEEL, 32 TEETH GEAR

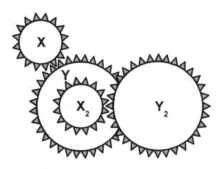

(b) PINION X$_2$, ATTACHED TO Y WHEEL, X$_2$ DRIVES Y$_2$

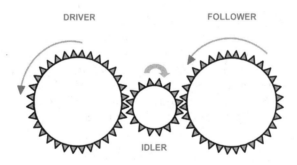

(c) IDLERS CHANGE ROTATION OF OUTPUT WHEEL, DRIVER & FOLLOWER ROTATE IN SAME DIRECTION.

FIGURE 4-4 (*a*) The pinion, gear X, with 16 teeth and the wheel, gear Y, with 32 teeth. (*b*) The pinion (gear X_2) is attached to X to drive the larger wheel (Y_2). (*c*) Gears with idlers to change the rotation of the output wheel to rotate in the same direction of the input wheel.

$R = 3$
$T_p = 10$ inch-pound
$e = 0.8$ (which is at 80 percent efficiency)
$S_p = 200$ rmp

Thus the output wheel torque will be

$$16 \text{ lb/in} = 0.8 \, (10 \times 2)$$

The speed of the wheel will be

$$80 \text{ rpm} = 0.8 \, (200/2)$$

If a higher gear reduction is needed, adding more gears won't do it, and in addition, the efficiency will drop. The reduction of any gear is a function of the gear ratio, the number of teeth on the output gear connected to the shaft (follower gear Y) divided by the number of teeth on the input gear connected to the motor (driver gear X). Adding gears between the driver and the follower, called *idlers*, will not cause an increase in the gear ratio. The ratio is between the same two gears. In order to have a larger gear ratio, gears can be layered. Attach a pinion (smaller gear X_2) to the wheel shaft Y, and use that second pinion to drive the larger wheel (Y_2), as shown in Figure 4-4b. As far as size, say that:

$$X = X_2$$

$$Y = Y_2$$

Then the gear ratio is shown. X_2 attached to Y rotates at the same speed. Therefore, X revolves four times (input rotations) for every one revolution of Y (output rotation), a 4:1 reduction. This is commonly done inside gearboxes. The gear efficiency should be supplied by the manufacturer of the gear on the data sheet based on the type of gear. Table 4-2 is a list of typical LEGO gears, their number of teeth and their estimated efficiency constant.

Calculating the total gear reduction of a gearbox or drive train is simple. Multiply all the gear ratios of the gear sets. For example, for

- Gear set 1: 4:1
- Gear set 2: 3:1
- Gear set 3: 6:1

the total gear reduction is $4 \times 3 \times 6 = 72{:}1$.

Thus, if a motor has a speed of 3,000 rpm with a gear ratio of 72:1, then the revolutions per minute for the motor will be $3{,}000/72 = 42$ rpm. Use the gear ratio to determine the effect on the torque of the motor.

From our previous list, using gears also can change the rotational direction. With two gears such as the wheel and pinion, the rotational output will be reversed. Thus, if

TABLE 4-2   List of Typical LEGO Gears

Gear Type	Description	# of Teeth	Estimated Gear Efficiency
Spur	Radial teeth (teeth that spread from the center outwards) parallel to the axis Most commonly used Used when shaft is in the same plane Have same size teeth so they can mesh properly	8, 16, 24, 40	~90% Highest Possible Efficiency
Bevel	Engages in another bevel gear on a different angle (usually 90 degrees) Used when the shafts to be turned meet at an angle; good for changing rotation angle Can only be meshed with another bevel gear Double-beveled gears have teeth beveled on both sides of the gear	12, 20, 36	~70% Low Efficiency
Worm	Has a short rotating screw that meshes with the teeth of another gear (usually a spur) Has a very high gearing ratio Not back-drivable, gravity or counter forces will not cause any rotation; can be used when a robot arm is holding something heavy so power will not be wasted on holding torque	No teeth	~70% Low Efficiency

the pinion, the input wheel, is turning clockwise, then the wheel, the output gear, will turn counterclockwise. However, if you want the output wheel to turn in the same direction as the input wheel, you will need to add an idler gear between them, as in Figure 4-4c. The output wheel now turns clockwise, the same as the input gear. What if you have a chain of, say, four or five gears touching? The rule is this: With an odd number of gears, rotation is in the same direction, and with an even number of gears, rotation is counterrotational. As far as the gear ratio is concerned, it stays the same. You are still calculating the input and output gears; all other idler gears are ignored. The gear efficiency is

$$\text{Total gear efficiency} = \text{gear type efficiency}^{\text{no. of gears} - 1}$$

Servo Gears = DC Motor + Gear Train

Servos are motors and gears inside a single plastic housing. The servo motor consists of a DC motor, a gear system, a rotation/position sensor, and some control electronics. The DC motor is very small. Servos typically have a range of movement of 60 degrees, but the Tetrix servos have 90 and 180 degree ranges. However, since

LEGO servos are used to rotate the wheels of our robots, they have a 360-degree range of movement clockwise and counterclockwise. A servo motor's output shaft can be moved to a specific angular position or rotation speed by sending it a coded signal. The servo motor will maintain the position of the shaft or speed as long as the same coded signal is applied. When the coded signal changes, then the shaft angle or speed will change too.

There are different types of servos. LEGO servos look nothing like Hitec servos. Hitec servos are RC servos. And LEGO servos, well, they have their own thing going on. But both have the same components. They are laid out quite differently, though. As with the shape and size of these motors, each has its advantages and disadvantages. Figure 4-5 shows (*a*) LEGO servo and (*b*) Hitec RC servo inside components.

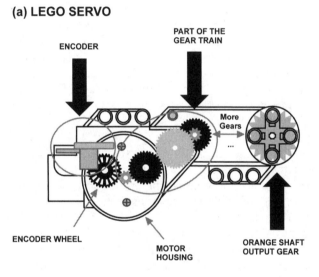

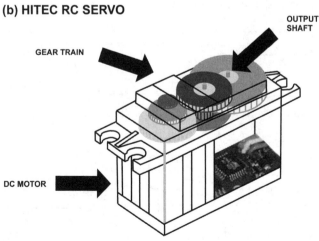

FIGURE 4-5   Inside components of (*a*) LEGO and (*b*) Hitec servos.

This is how a servo works: An electrical signal or pulse is sent to the motor. This is the commanded or new output for the shaft. For an NXT servo motor, this is the little orange thing where you can put the shaft in the middle. Servos also use pulse-proportional modulation (PPM), sometimes considered the PWM. The duration of the pulse (between 1 and 2 ms) determines the angular position to rotate to or the speed. The controller relays the PPM signal to the servo, which tells it where to go. With the Tetrix servo, the controller is external, but with the LEGO servo, the controller is internal. The position or speed is maintained until something new is signaled. The potentiometer measures the position of the output shaft at all times so that the controller board can correctly place and maintain the servo shaft at the desired position. The tachometer measures the speed of rotation. This is the job of the servo to maintain the desired output.

The servo uses closed-loop control, which feeds back to the controller so that adjustments can be made to obtain the desired position or speed. When the pulse changes, the error is determined between the current position/speed and the new position/speed. The servo motor will continuously make adjustments, resisting any movements in order to maintain the feedback operation that occurs every so many milliseconds. The electrical pulse is first fed to a voltage converter. The current rotational position or speed of the output shaft is read by the tachometer/pot, which produces a voltage. The tachometer serves as an electromechanical device that converts mechanical rotation into an electrical pulse. The rotational position is related to the absolute angle of the output shaft. Both signals, one representing the current position/speed and the other representing the new or commanded position/speed, are fed to an error amplifier. The output of the error amplifier is what drives the motor. The error amplifier determines the difference between the voltages. If the output is negative, the motor is moved in one direction, and if the output is positive, the motor is moved in another direction. The greater the difference, the greater the voltage, and the more the motor will move or turn. When the motor turns, this engages the gear system, which turns the output shaft to the commanded position. This process is depicted in Figure 4-6.

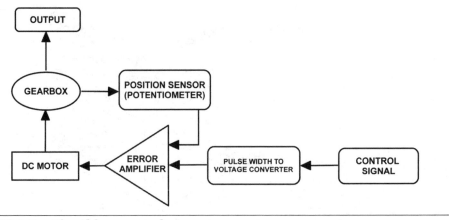

FIGURE 4-6   Closed-loop control.

The proportional integral derivative (PID) algorithm is at the core of closed-loop control. This algorithm is used to maintain speed so that power can be applied smoothly and quickly to bring the motor to its desired velocity. The algorithm goes something like this:

1. Read the current commanded set point (SP, desired output), the desired motor speed in revolutions per minute.
2. Read the process variable (PV, the feedback value returned in the process), the current motor speed read by the tachometer.
3. Subtract the PV from the SP to get the error signal (negative or positive). A positive value means that the motor is running too slow, and a negative value means that the motor is running too fast.
4. Use the PID function to calculate the control variable (CV).
5. If the error is nonzero, change it to a valid value for the CV (output of the control loop to the process), the new value for the PWM duty cycle.
6. Repeat.

Figure 4-7 shows (*a*) the controller loop and (*b*) the PID controller loop. Figure 4-7*c* provides an example of controlling the speed. PID gives your robot the ability to self-adjust its motor power levels to the correct values to maintain the desired speed. The closed-loop system monitors the error difference between how fast the robot is going and how fast it should be going and makes adjustments to the motor's power level, if necessary.

Here Come the Regulators: Encoders In and Out

When we first got our Tetrix robotics system, we were pretty excited about it. Tetrix has the aluminum beams, screws, gears, and those huge DC motors. We couldn't wait to build a robot and program it. So we built the basic chassis and added the robotic arm. We were ready to program those DC motors and see what they could do. We used the **TetrixMotor** class from the leJOS class library that only has a few methods to control the motor. This was the motor class to use when no encoders were installed. Even using low power, moving the robot forward caused it to speed across the floor. We needed encoders in order to have the type of control for the robot to be useful.

Encoders are devices that convert motion into a sequence of digital pulses. Encoders can be built into the motors, as in the case of LEGO servos, or they can be external, as in the case of the DC motor, where they are mounted on the outside of the motor. Figure 4-8 shows the location of the encoder for a LEGO servo and a DC motor. Hitec servos do not have encoders. The circuit control serves the purpose of monitoring the position of the shaft. Encoders can have a linear or rotary configuration, but the most common is the rotary type. There are two basic forms of rotary encoders: Absolute encoders have a unique digital word for each rotational

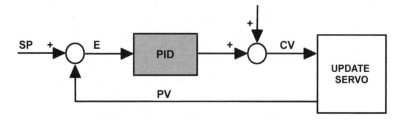

(a) CONTROLLER LOOP

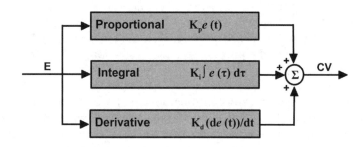

(b) PID CONTROLLER LOOP

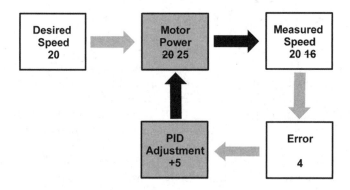

(c) EXAMPLE OF CONTROLLING SPEED

FIGURE 4-7   (*a*) Controller loop. (*b*) PID controller loop. (*c*) An example of controlling the speed.

LEGO BUILT-IN ENCODER AND WHEEL

TETRIX DC MOTOR'S WHEEL-MOUNTED EXTERNAL ENCODER

FIGURE 4-8 Location of the encoder for LEGO servos and the DC motor.

position of the shaft, and incremental encoders produce a digital pulse as the shaft rotates, allowing measurement of the relative position of the shaft. Most rotary encoders are made of a glass or plastic code disk. Some have a photographically deposited radial pattern organized in tracks. The NXT LEGO servo has a black wheel with 12 openings or slits. As the motor spins, the encoder wheel also spins, interrupting a beam of light on one side of the wheel that is detected by sensors on the other side of the wheel. The rotation generates a series of on and off states, as illustrated in Figure 4-9. Bits also can be combined to produce a single-byte output of n bits as a parallel transmission.

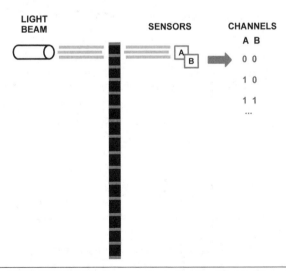

FIGURE 4-9 The NXT LEGO black encoder wheel with 12 slits generating on and off states.

Both the Tetrix encoder (sold by Pitsco) and the LEGO NXT servo encoder are quadrature optical incremental encoders. This means that the encoders will output a relative position and the direction of the motor. They use two sensors that are arranged along the encoder wheel in such a way that when one sensor encounters a black-white or open-closed boundary that causes a change of state, the other sensor will be exactly in the middle of a strip or open-closed slit, which means a steady state. The LEGO servo encoder wheel only has one level of slits, whereas other servos have two levels, with a sensor for each level. With one level, the sensors are arranged along that level on the wheel. The output for each sensor is 90 degrees out of phase with respect to the other sensor. Each sensor has its own channel, A and B. To determine speed, the velocity of the wheel, the accumulated pulse is sampled at fixed intervals. Figure 4-10 shows the output of the wheel encoder rotating counterclockwise when A leads B (*a*) and clockwise when B leads A (*b*).

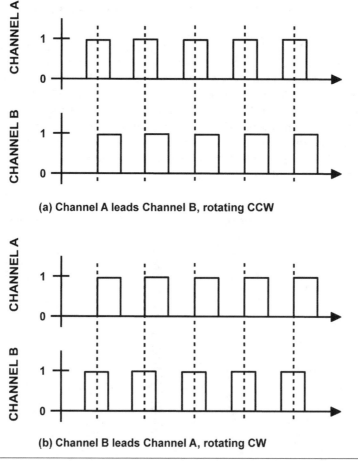

(a) Channel A leads Channel B, rotating CCW

(b) Channel B leads Channel A, rotating CW

FIGURE 4-10 The output of the wheel encoder rotating (*a*) counterclockwise when A leads B and (*b*) clockwise when B leads A.

Using Torque and Speed to Determine Selection of Team Members

How do we use these calculations for speed, torque, and whatever to choose which robot team members will perform a task? We can use the calculations of torque and speed to see which robots can safely perform the task at hand based on the capabilities of their motors. The amount of torque required by a robot to perform a task is called the *startup torque* and should be around 20 to 30 percent of its maximum torque. It should not exceed the rated torque for the motor. Say that a robot must accelerate to a speed of 30 cm/s; that is the target. You have a few robots to choose from to execute this task. One robot is small, weighs 1 kg, and has two wheels that are 3 cm in diameter (1.5-cm radius). A larger robot weighs 3.2 kg and has two wheels that are 7 cm in diameter (3.5-cm radius). The two robots have rated torques of 5 and 10 kg/cm and a maximum torque of 10 and 20 kg/cm, respectively. We can use a motion formula that relates the weight of the robot, torque, and acceleration:

$$C/r = m \times a + F_{attr}$$

where m is the mass of the robot (weight in kilograms), r is the radius of the wheel, a is the acceleration, and C is the torque. F_{attr} is the attrition force between the wheel and the floor. This has to do with the terrain challenges we discussed earlier and can be difficult to determine. Thus, for the sake of simplification, we will represent this value as $m \times a$. Thus:

$$C/r = 2(m \times a)$$

Assuming that the robot has two motors (left and right wheels), we calculate half this torque:

$$C = m \times a \times r$$

In Table 4-3 we can compare the torque and speed of these two robots and then make a decision on which robot will be best to perform the task.

As you can see, both Robot A and Robot B are far below their rated torque (maximum torque in which the motor will run continuously with no problems). That's fine. But the speed that is required for these robots far exceeds their revolutions per minute.

Summarizing DC and Servos Motors

Now that we have discussed how these motors work and some of their features, how do these motors compare? Which is the most powerful motor? What size robots would these motors be best at moving? Under what conditions? DC motors are the monster motors; they are really good for all types of tasks actually. Servos are good for precise

TABLE 4-3 Torque and Speed Comparison for Team Members, Robots A and B

Attribute	Robot A (A1R1)	Robot B (D1R1)
Weight m	1 kg	3.2 kg
Wheel radius r	1.5 cm	3.5 cm
Motor type	Servo	DC motor
No-load speed (rpm)	170 rpm	146 rpm
No-load torque (stall)	5 kg/cm	20 kg/cm
Rated torque	2.5 kg/cm	11.72 kg/cm
Desired acceleration a	0.80 m/s²	0.80 m/s²
Desired speed ds	80 cm/s	80 cm/s
$C = m \times a \times r$	$1 \times 0.8 \times 1.5 = 1.2$ N $\times$ m $= 0.12$ kg/cm	$3.2 \times 0.8 \times 3.5 = 8.96$ N $\times$ m $= 0.896$ kg/cm
Rpm calculation:	$6.28 \times 1.5 = 9.42$ cm	$6.28 \times 3.5 = 21.98$ cm
(1) Distance per revolution $= 2 \times 3.141 \times r$	80 cm/s	80 cm/s
(2) Rpm $= (ds/$distance per revolution$) \times 60$	$80/9.42 = 8.4 \times 60 = 509$ rpm	$80/21.98 = 3.6 \times 60 = 218$ rpm
Torque evaluation	0.12 kg/cm < 2.5 kg/cm	0.896 kg/cm < 11.72 kg/cm
Speed evaluation	509 rpm > 170 rpm	218 rpm > 146 rpm

movements such as those needed for robot arms and end-effector manipulation. Table 4-4 lists the features of each type of motor and some of their advantages and disadvantages.

Controlling the Motors: Tetrix Controller and NXT Brick

Controlling the motors, sending signals/commands for speed and positioning the shaft, is the responsibility of the controller, whether it is a microcontroller or just a simple dedicated DC motor or servo controller such as those for Tetrix. With LEGO MINDSTORMS, the microcontroller sends signals and commands to the servos. Figure 4-11 shows that the Tetrix robotics system has both microcontroller(s) and dedicated DC and servo controllers. The first controller of the daisy chain in our case is a DC motor controller plugged into sensor port 1. The controller acts as an intermediary between the

- Microcontroller

TABLE 4-4 Some of the Features, Advantages, and Disadvantages of DC and Servo Motors

Type of Motor	Feature	Weight Class	Advantages	Disadvantages
DC motors	Some are gearheads Use PWM duty cycle to reduce/control speed	Used for any weight robot	A lot of variety Very powerful Easy to interface Good for large robots	Too fast Require a high current Expensive Hard to mount on wheels Complex controls (PWM) Require an encoder
Servo	Has DC motor included Uses closed-loop control to maintain speed/position	Robots up to 5 pounds	A lot of variety Good for indoor robot speed, small robots Cheap Easy to mount on wheels Easy interface Medium power requirement	Low weight capabilities Little speed control

- Batteries
- Motors

A motor controller may be needed because the microcontroller usually can only provide roughly 0.1 A of current, whereas most actuators such as motors require several amperes. Another advantage is that instead of using up the output ports on the NXT brick, only one input port is used. This is possible with the I^2C protocol. With one DC motor and servo controller, two DC motors with two encoders, and up to six servos can be controlled.

These controllers can be daisy-chained (connecting multiple controllers together), allowing for many motors (servo and DC motors) to be used by one robot. Up to four controllers can be daisy-chained and connected to a single sensor port. Daisy-chaining works by having the first controller in the chain read all the messages sent out by the NXT. The controller passes along all the messages that are not intended for that specific controller. The messages are passed sequentially along down the chain until they reach the intended controller. Each controller is assigned a unique address (the scheme is based on the software used for development) to ensure that the messages sent out get to the correct controller. If the controllers are not configured properly, then the messages that are sent out to the controllers will end up at the wrong controller and the wrong motor, and the robot will behave strangely. Figure 4-12 shows daisy-chained DC controllers and the motors connected to them.

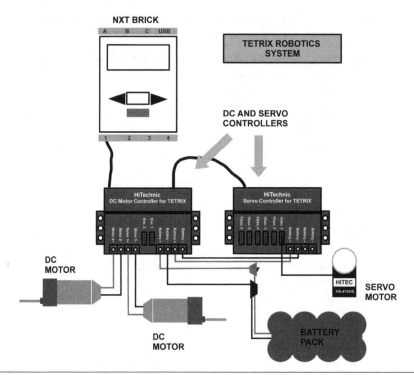

FIGURE 4-11 Tetrix robotics system with both microcontroller(s) and dedicated DC and servo controllers.

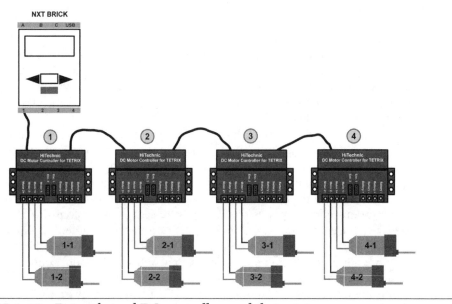

FIGURE 4-12 Daisy-chained DC controllers and their motors.

Using the Motors

leJOS has several classes to be used to program all the DC and servo motors and DC and servo controllers. Table 4-5 lists the classes and methods with brief descriptions of the motor classes.

TABLE 4-5   leJOS Classes and Methods for Motors and Controllers

Controllers	
`TetrixMotorController` class	HiTechnic Tetrix motor controller abstraction; provides `TetrixMotor` and `TetrixEncoderMotor` with instances used to control the Tetrix motors
Methods	`TetrixMotor getBasicMotor(int motorID)`
	`TetrixEncoderMotor getEncoderMotor(int motorID)`
	`TetrixRegulatedMotor getRegulatedMotor(int motorID)`
	`float getVoltage()`
`TetrixServoMotorController` class	HiTechnic servo controller abstraction; provides `TetrixServo` instances that are used to control Tetrix servos
Methods	`void flt()`
	`TetrixServo getServo(int servoID)`
	`int getStepTime()`
	`boolean isMoving()`
	`void setStepTime(int step)`
DC Motor Classes	
`TetrixMotor` class	Tetrix basic DC motor abstraction without encoder support
Methods	`void backward()`
	`void flt()`
	`void forward()`
	`int getPower()`
	`boolean isMoving()`
	`void setPower(int power)`
	`void setReverse(boolean reverse)`
	`void stop()`
`TetrixEncoderMotor` class	Tetrix DC motor abstraction with encoder support

(continued on next page)

TABLE 4-5   leJOS Classes and Methods for Motors and Controllers (*continued*)

Methods	`int getLimitAngle()`
	`int getTachoCount()`
	`void resetTachoCount()`
	`void rotate(int degrees, boolean immediateReturn)`
	`void rotateTo(int limitAngle, boolean immediateReturn)`
	`void setRegulate(boolean regulate)`
`TetrixRegulatedMotor class`	Tetrix DC motor abstraction with encoder support that implements `RegulatedMotor`
Methods	`void addListener (RegulatedMotorListener listener)`
	`void backward()`
	`void flt(boolean immediateReturn)`
	`void forward()`
	`float getMaxSpeed()`
	`int gerRotationalSpeed()`
	`int getSpeed()`
	`boolean isStalled()`
	`RegulatedMotorListener removeListener()`
	`void rotate(int angle)`
	`void rotate(int degrees, boolean immediateReturn)`
	`void rotate(int limitAngle)`
	`void rotate(int limitAngle, boolean immediateReturn)`
	`void setAcceleration(int acceleration)`
	`void setRegulate(boolean regulate)`
	`void setSpeed(int speed)`
	`void setStallThreshold(int error, int time) void stop(boolean immediateReturn)`
	`void waitComplete()`
Servo Motor Classes	
`TetrixServo class`	Basic servo motor abstraction

TABLE 4-5   leJOS Classes and Methods for Motors and Controllers (*continued*)

Methods	`float getAngle()`
	`int getpulseWidth()`
	`void setAngle(float angle)`
	`void setpulseWidth(int microSeconds)`
	`void setRange(int microsecLOW, int microsecHIGH, int travelRange)`

The motor that is being programmed has to be identified. In the constructor for the controller classes, `TetrixServoController` and `TetrixController`, you have to identify:

- `port`
- `daisyChainPosition`

`port` is the sensor port to which the controller is connected. If it is daisy-chained, it is the first controller's sensor port. `daisyChainPosition` is the position of the controller in the daisy chain. Some methods require the motor ID. This is an integer and refers to the channel into which the motor is plugged on the controller, `MOTOR_1` or `MOTOR_2` on the DC motor controller channels 1 and 2, respectively. For servo motors, there can be up to six servos, so there are six channels on the controller. The servo `motorID` value stands for `SERVO_1` through `SERVO_6`. There are two classes that define a DC motor with encoder support, `TetrixRegulatedMotor` and `TetrixEncoderMotor`. The class `TetrixMotor` is defined for a DC motor that does not have encoder support.

LabVIEW defines blocks for both Tetrix DC and servo motors and LEGO servo motors. LEGO servo blocks are located in the Functions palette under "NXT Native I/O | Output" options. There you will find a bunch of motor blocks, as shown in Figure 4-13a. These blocks will program the servo motors, individually or in sync, to move, turn, stop, move an unlimited direction, move a fixed distance, or move for a fixed time at a specified speed and power level.

The Tetrix DC and servo motor blocks are located in the Functions palette under the "NXT Robotics | TETRIX" options (Figure 4-13b). These blocks will program the DC motors to move, move a fixed distance, check status, and stop. There are blocks to move and check the servo motors. There is also a block to reset encoders. The use of these blocks is pretty straightforward. To program the motors, DC or servo, you must specify which motors you are programming with the motor ID number, just as you do when using leJOS. The motor ID is a three-digit number; the first number is the port, the second number is the controller number (position in the chain), and the last number is the channel on the controller. Figure 4-14 shows that when creating

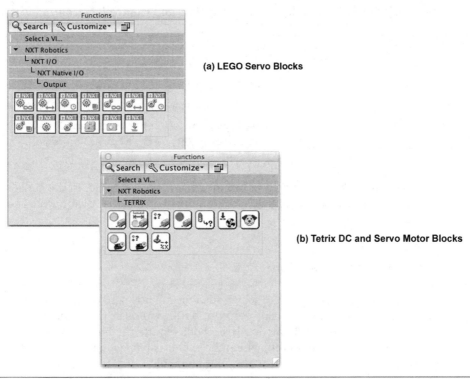

FIGURE 4-13 LabVIEW's function palette for the (*a*) LEGO servo blocks and (*b*) Tetrix DC and servo motor blocks.

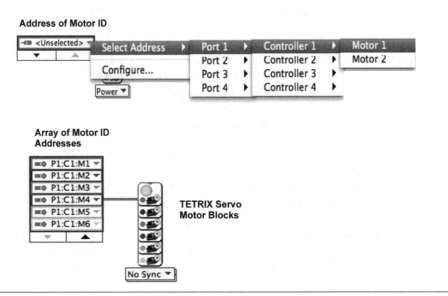

FIGURE 4-14 Tetrix motor block and array of motor ID constants.

a constant, a box labeled "UNSELECTED" appears. This is where the motor number goes. Click on the triangle "SELECT ADDRESS" option, and choose the correct port, controller, and channel number. Multiple motor ID numbers are stored here for an array of motor numbers.

To use a motor equipped with an encoder, use the "Move a fixed distance" block. With this block, the specified distance is expressed as encoders counts. This is an integer; if it is positive, then the motor turns clockwise, and well, you know the rest. To calculate the encoder's count, convert the distance, in inches, to encoder counts:

1. Calculate the circumference: Circumference = diameter × 3.141.
2. Convert the travel distance in inches: Rotations = travel distance/ circumference.
3. Convert rotations to total degrees: Total degrees = 360 × rotations.
4. Convert total degrees to encoder counts, which are one-quarter of a degree: 4 × total degrees.

Figure 4-15 shows this calculation performed in LabVIEW. The result is then inputted to the distance of the block.

NXT-G defines blocks for LEGO NXT servos. Figure 4-16 shows the NXT Rotation (*a*), Move (*b*), and Motor (*c*) blocks. The Rotation block counts the number of degrees the motor connected to a specific port rotates. One full rotation is 360 degrees. By using the data wires, this block can send out the current number of degrees or rotations and a logic signal (Boolean true/false) based on whether the number of degrees or rotations falls above or below some set trigger point.

The other two blocks are the Move and Motor blocks. The simplest way to tell the difference between these blocks is by the color of the background stripe and the number of gears on the block. Motor blocks have a light-green stripe with a picture of one large gear. Move blocks have a dark-green stripe with a picture of two gears, one

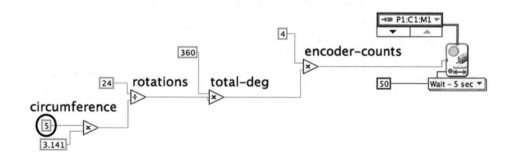

NOTE: Replace with Diameter of the wheel

FIGURE 4-15   Calculation for encoder counts in LabVIEW.

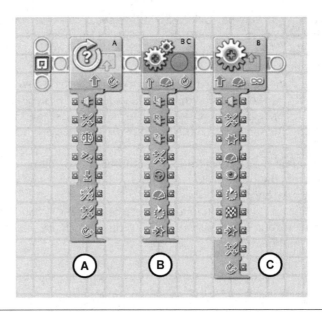

FIGURE 4-16 NXT-G blocks for LEGO NXT servos (*a*) rotation, (*b*) move, and (*c*) motor blocks.

large and one small. Use the Move block to move forward or backward in a straight line or to turn by following a curve. Use the Duration property to set how far the robot will go. Use the Motor block for precise control of the motor's speed. Use **Ramp_up** to a set speed or **Ramp_down** to stop the motor. Deselect the "Wait for Completion" option so that the program will move to the next block once the Motor block has started a motor, but if you want the motor to complete, then leave the option checked. Table 4-6 lists the advantages and disadvantages of using the Move and Motor blocks.

NXT-G PID Block

There is a NXT-G block from HiTechnic that implements the proportional integral derivative (PID) algorithm for the Motor block. It is considered to be a Motor block. The standard Motor block uses relative rotation because the relative rotation encoder is built into the servo, as you know. This block will give absolute rotation control over the motor. With this implementation, the set point (SP) is the desired angle position. Thus, if you want the motor to move to a desired location, the location or angle is derived from the reference angle or position, and it will advance from there, not the last position of the motor. The PID block also has a few reset actions or a few ways to establish the reference position (zero position) of the motor. This block is good for

TABLE 4-6 Advantages and Disadvantages of Move and Motor Blocks

Block Type	Advantages	Disadvantages
Move block	Some routine tasks are done automatically; this will eliminate the need for extra commands, such as using the rotation sensor to reset the motor; it also makes correcting the course straightforward when drifting to one side. Move blocks programs take less memory to store on the NXT brick. Will take fewer blocks for the robot to perform the same behaviors; this will save screen space and makes programs easier to read.	Built-in functions designed to improve robot's performance will sometimes cause robot to act strangely when something unexpected happens to the robot; for example, the motor plugged into the wrong port will cause the motor in the correct port to freeze. Different patterns must be learned to perform different behaviors: How many blocks are needed to perform the behavior? When making modifications, advantages are not used with nonstandard designs.
Motor block	Easy to troubleshoot; all actions are explicit, not work done automatically. Blocks always work the same regardless of robot design. Consistent five-block pattern used in almost all types of robot movements.	Does not correct motor when drifting. Takes more blocks to perform a single movement. Have to manually reset the rotation sensor between measured movements. Programs take up more space and more memory space on the NXT brick.

when you want to advance the motor to a certain position relative to the reference point. This is good when you do not know what position the motor is currently in. You can download this block from the HiTechnic website. It is to be used instead of the Motor block. Figure 4-17a shows the block and the configuration panel. Figure 4-17b shows the difference between the relative positioning and the absolute positioning for a robot arm. This is an excellent Motor block and can be an improvement on programming a robotic arm.

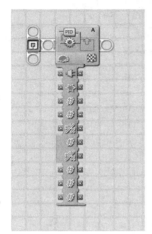
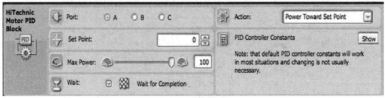

(a) PID MOTOR BLOCK AND CONFIGURATION PANEL

(b) RELATIVE AND ABSOLUTE POSITIONING OF ROBOT ARM

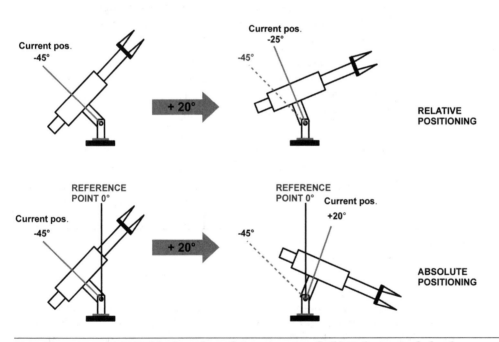

FIGURE 4-17 (*a*) PID block and configuration panel. (*b*) Relative and absolute positioning for a robot arm.

Robotic Arms and End Effectors

Motors are also used for arms and end effectors. A robot team may require different types of arms and end effectors to do a specific task or to be more autonomous in uncontrolled or controlled environments. The more diverse the team, the greater likelihood that there will be capable team members to handle unplanned or unforeseen situations they may encounter.

In our CSI Bluetooth Robotic-Oriented Network (BRON), we needed different types of sensors, such as pH, temperature, RFID, and ultrasonic sensors, to patrol the warehouse, identify items in the warehouse, determine whether there was something that did not belong, attempt to identify the substance, and so on. Once we identified a beaker that should not be there, we needed to call in our special unit that transported the beaker to a safe location so that it could be analyzed. We had two robots that potentially could be used for this task. Both were equipped with different types of robot arms and end effectors (Figure 4-18). Which one has the torque to lift and hold the beaker?

Robot Arms of Different Types

A robot arm is a programmable mechanical device that works like a human arm. Like a human arm, a robot arm has a length, joints, and a hand or end effector. It can be attached to a fully functioning robot (not necessarily bipedal) or work as a standalone device that has mobility or not. There are different types of robot arms. Figure 4-19 shows three basic robot arm types.

 There are LEGO MINDSTORMS NXT versions of each type of arm implemented by enthusiasts. You can do a search or see links to some of the best we have found on our website.

FIGURE 4-18   Two robots equipped with different types of robot arms and end effectors.

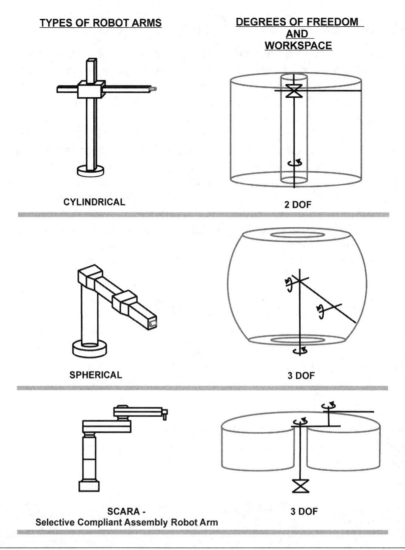

FIGURE 4-19 Three basic robot arm types.

Each type of arm is characterized by its

- Configuration space
- Workspace

as depicted in the figure. The *configuration space* of a robot arm is determined by the limitations of the robot arm's *degrees of freedom* (DOF). The degrees of freedom determines where a joint in the arm can "bend." The bend can be a rotation or a translation motion working in a three-axes plane (x, y, z). Translational motion involves moving the joint forward or backward. Figure 4-20*a* shows the rotational and translational motions of a robot arm. As for the two robots C and B, robot C has two controllable or programmable degrees of freedom, one for the end effector and the other for the arm but robot B has one controllable degree of freedom for the arm. To open and close the end effector, the arm is raised and lowered. Movement is transferred to perpendicular knob wheels that open and close the claw. Figure 4-20*b* shows the DOFs for both robot arms. For these robot arms to be considered spherical-type robot arms, the base would need to turn 360 degrees.

Each rotary joint has restrictions or limitations to its movements. These limitations can be based on the maximum angle for the servo. For example, Hitec servos are 90- and 180-degree servos, whereas LEGO servos are 360-degree servos. Although the servo has a maximum angle, the arm may use only a smaller degree owing to the design or location of the arm. The translational motions are forward, backward, vertical, and horizontal. There can be many degrees of freedom in a robot arm. The sum of all configuration spaces for all the joints of the arm, the link lengths, the angle in which objects can be lifted, and so on define the workspace. The workspace is the reachable space of the end effector in three-dimensional

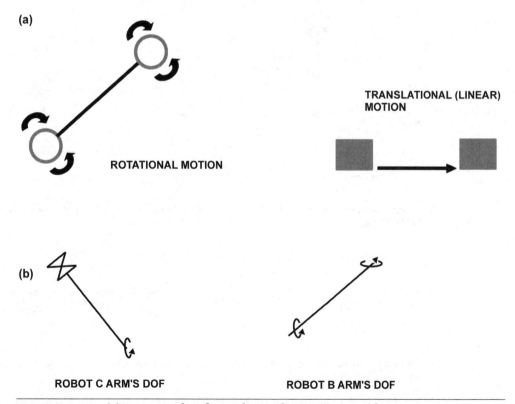

FIGURE 4-20   (*a*) Rotational and translational motions in a robot arm.
(*b*) The degrees of freedom of the robot arms for Robots C (C5R2) and B (D1R1).

(3D) space. It determines the fixed boundaries of the possible locations of the end effector. Locations outside the workspace are places that the arm and end effector are not capable of reaching. If the robot is mobile, this adds to the workspace. Its total workspace would include the area to which the robot could physically relocate. *Forward kinematics* is the method for figuring out the orientation and position of the end effector given the link lengths and joint angles. A discussion of forward kinematics is outside the scope of this book.

Torque of the Robot Arm

We have two robots (C and B) equipped with robot arms. We can calculate the torque of one of the robot arms when lifting an object. We want to know whether the robot arm is capable of lifting an object based on the torque output of the motors. If the determined torque value exceeds the maximum torque, then we know that the arm is unable to lift the object. In this case we are calculating the torque as it relates to the force acting on the length of the link to the pivot point of the arm. Looking at Figure 4-21a, we can see that the length is perpendicular from the pivot to the force. The

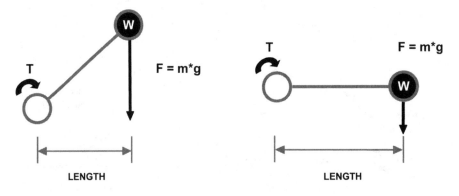

(a) Length Perpendicular from the Pivot to the Force.

(b) Fully extended Length when Torque is highest.

(c) Calculation of Torque includes: weight of object (W), actuators (A1 – A3), and mass of links (M1 – M3).

FIGURE 4-21 (*a*) Length is perpendicular from pivot to force. (*b*) Fully extended length when torque is at its highest. (*c*) Inclusion of object, arm, and actuator weights in the calculation.

force working on the end will be the weight of the object. The calculation we will use follows:

$$T = m \times g \times L$$

where $m \times g$ is the weight of the object, and L is the length. The length will be the worst case, which is the longest length to the pivot when the arm (or link) is fully extended (Figure 4-21b). This is also when torque use is at the highest. The calculation does not just include the weight of the object but also the weight of the arm and the actuators if significant (Figure 4-21c):

$$T_{arm} = L_{arm} \times W_{obj} + \frac{1}{2}L_{arm} \times W_{arm}$$

where L_{arm} is the length of the arm multiplied by the weight of the object W_{obj} added to one-half the length of the arm (center of mass) times the weight of the arm W_{arm}. It gets crazy very quickly with every pivot or joint in the robot arm. So we used an online robot arm torque calculator (www.robotshop.com/robot-arm.html). This calculator also includes the mass of each actuator in its calculation. Table 4-7 shows the values for the variables in this calculation for Robots C and B. It does not include the end effectors of the robots. Based on this calculation, the torque requirement to lift the beaker for Robot C will exceed its maximum torque where the torque requirement to lift the beaker for Robot B will not exceed its maximum torque.

TABLE 4-7   Calculation Results for the Torque for the Robot Arms for Robots C and B

Attribute	Robot C (C5R3)	Robot B (D1R1)
DOF (without end effector)	1	1
Length of arm L_{arm}	20 cm	16 cm
Weight of beaker with substance W_{obj}	.297 kg	.297 kg
Weight of arm W_{arm}	.061 kg	.145 kg
Stall torque (including additional gear sets)	5.098 kg cm	6.0 kg/cm
$T_{arm} = L_{arm} \times W_{obj} + \frac{1}{2}L_{arm} \times W_{arm}$	20 cm $\times$ .297 kg + $\frac{1}{2}$(20 cm) $\times$ .061 kg	16 cm $\times$ .297 kg + $\frac{1}{2}$(16 cm) $\times$ .145 kg
Torque comparison	6.55 kg cm > 5.098 kg cm	5.91 kg cm < 6.0 kg cm

End Effectors of Different Types

There are also many types of end effectors:

- Mechanical grippers
- Negative-pressure (vacuum) grippers
- Magnetic grippers
- Hooks
- Ladles (scoops liquid or powder)
- Others (electrostatic)

We use mechanical end effectors for our robot arm. These are the most common. Mechanical grippers can be classified as:

- Parallel grippers
- Angular grippers
- Toggle grippers

Figure 4-22a shows a few mechanical end effectors. In the figure you also can see the end effectors for the arms on Robots C and B in Figure 4-22b. They are angular with encompassing grippers.

To consider whether the end effectors of your robots can be used to transport our beaker, we can perform a similar calculation to the torque of the arm to determine whether the end effector has the torque to hold the beaker (as opposed to the arm's ability to lift the beaker):

$$F = \mu \times W_{obj} \times n$$

where F is the force required to hold the object, μ is the coefficient of friction, W_{obj} is the weight of the object, and n is the number of fingers on the gripper. Robot B uses a servo motor, but the gripper for Robot C is geared to the motor of the arm by means of a gear and linear displacement.

The torque of the gripper is not the sole determining factor. We also have to consider other aspects of the gripper. Gripping implies there are mechanical fingers of some type. There may be two, three, or five fingers, like a human hand. The number of fingers is actually considered in the preceding calculation. Our robots have two grippers, but these grippers are extremely different. What about the shape of the fingers? For Robot C, there is a roundness, whereas Robot B's end effector is square. A negative impression of the shape of the object (round beaker, square gripper) may be more efficient. Figure 4-23 shows both end effectors attempting to grip a beaker.

Software Support of the Robot Arm

As we discussed earlier, it is the duration of the pulse that determines the rotation of the servo motor. The leJOS Tetrix servo class Set Pulse Width can be used to set

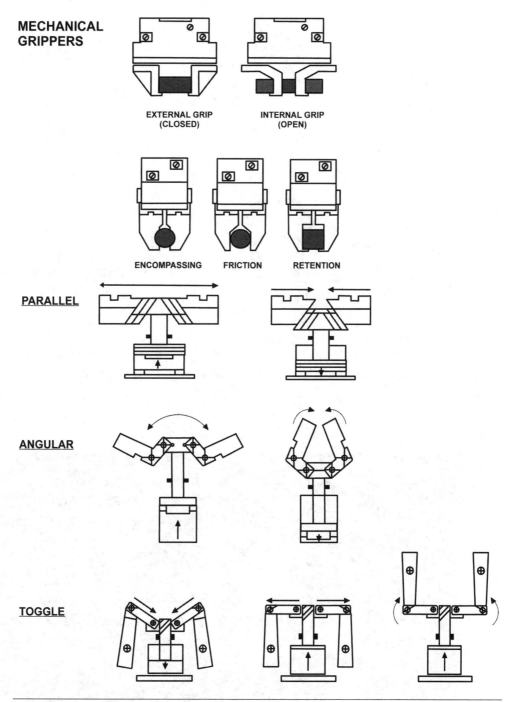

MECHANICAL GRIPPERS

EXTERNAL GRIP (CLOSED)

INTERNAL GRIP (OPEN)

ENCOMPASSING   FRICTION   RETENTION

PARALLEL

ANGULAR

TOGGLE

FIGURE 4-22   (a) Types of mechanical end effectors.

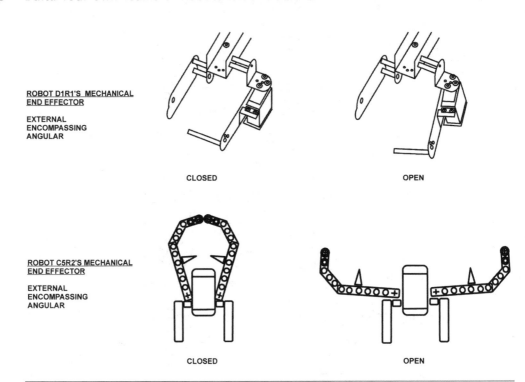

ROBOT D1R1'S MECHANICAL END EFFECTOR

EXTERNAL
ENCOMPASSING
ANGULAR

CLOSED   OPEN

ROBOT C5R2'S MECHANICAL END EFFECTOR

EXTERNAL
ENCOMPASSING
ANGULAR

CLOSED   OPEN

FIGURE 4-22   (*b*) End effectors for the arms on Robots C and B.

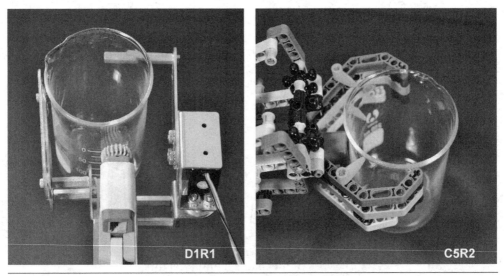

D1R1   C5R2

FIGURE 4-23   End effectors of Robot B (*left*) and C (*right*) gripping a beaker.

the position of the robot arm. The width of the pulse must be within the maximum angle range of the servo. This method uses an absolute, not a relative, pulse width in microseconds. The method accepts an integer that represents the microseconds. A "neutral" position is a width of 1,500 ms; this is the midpoint of the range of the servo. The parameter is between 750 and 2,250 ms with a step resolution of 5.88 ms. Also use these methods from the leJOS class library

- `setRange`
- `setAngle`

The `setRange` method sets the allowable pulse width within the range of the servo in microseconds and the total travel range. The method sets the `microsecLOW`, `microsecHIGH`, and `travelRange`. For example, 180 or 90 degrees would be the total travel range for Hitec servos. The default total range is 200 degrees, with the low at 750 and the high at 2,250 ms. This information must reflect the actual empirical specifications of a specific servo to be able to position the servo accurately.

Are the Three Laws of Robotics Real?

BRON'S BELIEVE IT OR NOT!

Law 1. A robot may not injure a human being or through inaction cause a human being to come to harm.

Law 2. A robot must obey orders given to it by human beings except when such order would conflict with the first law.

Law 3. A robot must protect its own existence as long as such protection does not conflict with the first and second laws.

Isaac Asimov introduced the three laws of robotics in 1942 in his short story "Runaround." The question is, Did Asimov simply invent these three laws for the purposes of the story, or were they a product of his careful consideration of how robots should interact with human beings? Were the three laws Isaac Asimov's way of warning us to never build a machine to which these three laws might legitimately apply? Believe it or not, there is much philosophical discussion of the three laws of robotics and how they could be really used in the interaction of robots and human beings. But! Stop the presses!

We build robots! We'll build several in this book. What sensors could we use to allow our robots to even identify a human being, let alone determine whether one of our robot's action were causing harm? How would our robot know that through its inaction some human being somewhere was being harmed? How do we get our robots to understand what a human being is in the first place? How would a robot know whether executing certain instructions would cause it to injure a human being? Anyway, what kind of injury are we talking about: physical injury or psychological injury? And to make matters worse, how do our robots know that they exist? The third law says that a robot must protect its own existence. We've programmed our robots not to go off edges, but is that the same thing as protecting their own existence? How would our robot know that if it refrains from going over an edge to its ultimate doom, it might be violating the second or third law?

It's not the reality of the three laws of robotics that we should be questioning. It's the feasibility of the type of robot that the three laws mention. Such a robot would have to have incredible powers of perception, cognition, foresight, contemplation, reasoning, and so on. It's not clear whether a robot like the one mentioned in the three laws could ever exist. As humans, it's possible for us to cause injury and not even be aware that we are causing it. Who could know that of all the things that we don't do, some of those things might cause injury as a result of us not doing them! No! We don't need to challenge the validity of the three laws. But the robots those laws refer to have never been built and, based on what we know about robots, could never be built.

Chapter 5

Bluetooth Programming in NXT-G and LabVIEW

The Lost Scrolls of Robotics: #5

I am not a gun.

— Iron Giant, *Iron Giant*

NXT-G is a graphical programming language for MINDSTORMS NXT based on LabVIEW and the G language (yes, the G in NXT-G). It was developed by LEGO and National Instruments (NI) to "streamline" and "make simple" but also keep some of the programming power of LabVIEW. NXT-G is meant to appeal to school children with no experience programming or interacting with robots. But the LEGO MINDSTORMS NXT platform, NXT-G, and LabVIEW for LEGO MINDSTORMS are also being used by university students.

In this chapter we will discuss the Bluetooth communication blocks in the NXT-G and LabVIEW for MINDSTORMS environments. In Chapter 2 we introduced Bluetooth communications for the MINDSTORMS NXT Bluetooth Robotic-Oriented Network (BRON). We discussed the nature of the Bluetooth signal and dispelled some myths. We covered communications protocols, NXT software layers for Bluetooth implementation, and introduced the basics of sending/receiving, writing/ reading Bluetooth communications between NXT bricks in the NXT-G and LabVIEW environments. Here we will go into some detail on the Bluetooth connection and communication blocks for these environments.

A Little Background Block by Block

LabVIEW is a system design development environment used by scientists and engineers to visually program and design devices. It is commonly used to acquire data from devices and for instrument control. LabVIEW for LEGO MINDSTORMS was developed for programming the NXT brick. LabVIEW, which stands for *Laboratory*

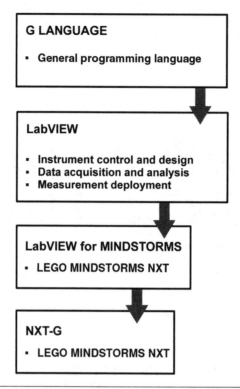

FIGURE 5-1   The relationship between G language, LabVIEW, and NXT-G environments.

Virtual Instrumentation Engineering Workbench, was developed by NI. It is based on G, a general-purpose programming language used for data acquisition, serial instrument control, data analysis, data presentation, and data storage. Figure 5-1 shows the relationship between G language, LabVIEW, and NXT-G. LabVIEW is not a text-based programming language but a graphical programming language. It uses icons and graphical blocks to represent programming elements or nodes such as loops, functions, structures, operations, and so on. A program called a *virtual instrument* (VI) is comprised of these graphical blocks and icons, which are connected by drawing wires that propagate data, called the *dataflow,* assembled in a block diagram. With dataflow programming, a node executes when all data inputs arrive. When the node finishes executing, all data are supplied to all of its output terminals. LabVIEW is the preferred programming tool by engineers because the actual appearance and operation of the program simulates the wiring of physical instruments.

LabVIEW (referring to LabVIEW for LEGO MINDSTORMS) uses a component approach to software development. The program that you are to build for your robots should be broken down into tasks and subtasks. The program or VIs are hierarchical and modular. Once a VI has been encapsulated in a block and included in other VIs, it is considered a subVI. SubVIs allow details to be hidden and for the subtask to be reused. The assembly of the VI or subVI, the source code, is constructed on the block diagram. The inputs and outputs are displayed as controls and indicators in the front panel serving as the interaction with the VI. Each control or indicator corresponds with an input or output terminal (where data are passed to or out of a VI), shown as a circle and a wire on the graphical block in the block diagram. The color determines the type of data. Figure 5-2 shows an example of a subVI (a simple calculator), the block diagram, front panel, and its block. The block diagram reflects the function model of input, process, and output. This is one of the example subVIs provided by NI in the "Examples" directory of LabVIEW.

The input wires protrude generally from the left side, and the output data wires protrude from the right side of the block and are labeled. Optional inputs are labeled and are gray. The default values are enclosed in parentheses. In NXT-G, the options for the block are set in the configuration panel. Input and output data wires are connected to a data hub that appears at the bottom of the block. Data wires connect the blocks, causing the block to execute as soon as the data are available in both environments. Movement of data along the data wires determines the next execution of the program. In LabVIEW, one block connects to a specific location on the block,

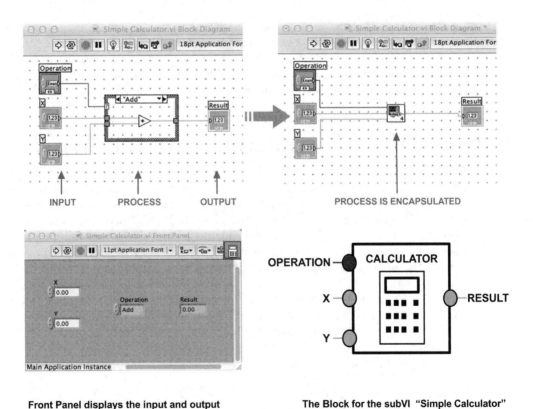

Front Panel displays the input and output
of the subVI "Simple Calculator."

The Block for the subVI "Simple Calculator"
to be used in other VIs.

FIGURE 5-2   An example of a subVI, the block diagram, and front panel.

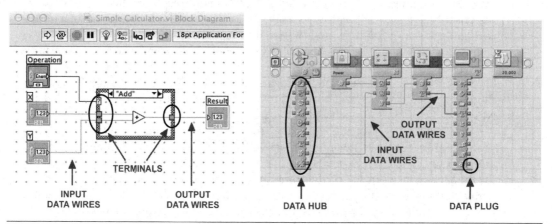

FIGURE 5-3 Connections of blocks using data wires in (*a*) LabVIEW and (*b*) NXT-G.

whereas in NXT-G, program sequences are based on the sequence beam. Figure 5-3 shows how data wires connect blocks in LabVIEW and NXT-G.

These subVIs define blocks for things; for example, the functionality of objects such as sensors and motors, statements, and operations such as data conversions and mathematical and Bluetooth operations.

Establishing a Connection with the BRON

Before discussing the Bluetooth blocks for the NXT-G and LabVIEW environments, let's recap the process of establishing the connection with all the NXT robots that will be members of the BRON. This process is identical for each environment because the Bluetooth connections are established between the NXT bricks. Bluetooth communication requires that the devices identify themselves to one another or, more specifically, that all members must identify themselves and exchange passkeys with the initiator/team leader in order to have permission to communicate. This is called *pairing* with the NXT bricks. In Chapter 2 we describe the basic steps of pairing D1R2 and D1R1, where D1R2 is the team leader. Below are the basic steps of connecting NXT-based robots for a BRON (all bricks for the NXT-based robots that will be members of the BRON should be on; Bluetooth capability also should be on, as well as visibility).

From the brick of the team leader:

- Select the Bluetooth submenu.
- Select the "Search" option in order to perform the scan for other NXT bricks.
- Select the first NXT brick of the team member from the resulting list.
- Select the connection line for that brick.

- If this is the first time connecting with that brick, select the default passkey (if used) or enter the passkey for that device.
- Confirm the connection with the team member by entering the same passkey on its brick.

Go through the same process for as many bricks that are members of the BRON (up to three).

Once the initial connections are made, entering the passkey for the brick will no longer be necessary. The bricks with which the team leader has connected will be listed in its "My Contacts" list. Whenever a BRON communication is to be constructed, the "Search" should be performed in order to determine which bricks are available (they may be off, out of range, not visible, etc.). The connection line has be selected each time. If a new team leader is chosen, this process has to be repeated on the new team leader's brick. Once the initial connection has been made, Bluetooth communication can be used in the LabVIEW or NXT-G environments. Make sure that you remember the connection number for each NXT brick. This is the connection number the team leader uses when sending and receiving messages to and from the team members.

The NXT brick displays an indicator in the upper left corner that signifies

- Bluetooth capability is off.
- Bluetooth capability is on and visible.
- Bluetooth capability is on, and the brick is visible and connected to another Bluetooth device.

This is shown is Figure 5-4.

Connecting a PC to NXT Bricks from NXT-G and LabVIEW

From NXT-G, make sure that all team members of the BRON are available to your computer via Bluetooth or USB. This is the same process as connecting from the NXT bricks. The "NXT Window" option on the controller (upper-left button) will list the

✳	**Bluetooth is On but the NXT Brick is not visible.**
✳‹	**Bluetooth is On and the NXT Brick is visible.**
✳◇	**Bluetooth is On and the NXT Brick is connected.**

No indicator means Bluetooth is Off.

FIGURE 5-4   The Bluetooth indicators displayed on the brick's screen.

names of the devices. The list will include the NXT bricks to which you can connect. For each device, the window lists these attributes:

- Connection type (Bluetooth or USB)
- Status

The status of the device will be

- Unavailable
- Available
- Connected

If the NXT brick's status is "Unavailable," it cannot be connected to—it is not seen by the computer. If the NXT brick has a USB connection type, this means that the device is currently not tethered to the computer using the USB cable or that the device is off. If the device is a Bluetooth connection type, the NXT brick may be off, not visible, out of range, or not paired with the computer. If the device is "Available," it is seen and can be connected to. A "Connected" status means that the device is currently connected to the computer. This becomes the targeted NXT brick to which the computer will download the programs.

The scan function will search for all visible Bluetooth devices within the scan area. Select the NXT brick that is to be a member of the BRON. You will be asked for the passkey if this is the first time connecting. Confirm the passkey on the NXT brick to which you are pairing. Only one brick will be connected to your computer at a time via Bluetooth or USB. Pair with each NXT brick that is in the BRON. Figure 5-5 shows the steps for connecting a new NXT brick initiated from the "NXT Window" on a Mac.

The NXT brick with a status of "Connected" is the context for NXT-G programming. When switching to another brick, you will have to select that NXT brick from the "NXT Window" and connect to it. Make sure that you are aware of which brick is the team leader.

The process of connecting the NXT brick to the computer from LabVIEW is identical to the process in NXT-G. From the NXT terminal, select "Find NXT." The process is the same as the process of scanning, pairing, and connecting to the new NXT brick. In LabVIEW, once the NXT bricks have been scanned, the NXT terminal dialog will list the NXT bricks. The dialog will display the status of a selected brick (whether it has a Bluetooth or USB connection type) as being "Connected" or "Not connected." Figure 5-6 shows the NXT terminal listing the NXT brick and the status of a selected brick.

Connecting to the BRON

Once the BRON has been established, Bluetooth communication blocks can be included in the NXT programs. Here are the Bluetooth blocks:

- On/Off
- Connection/Disconnection

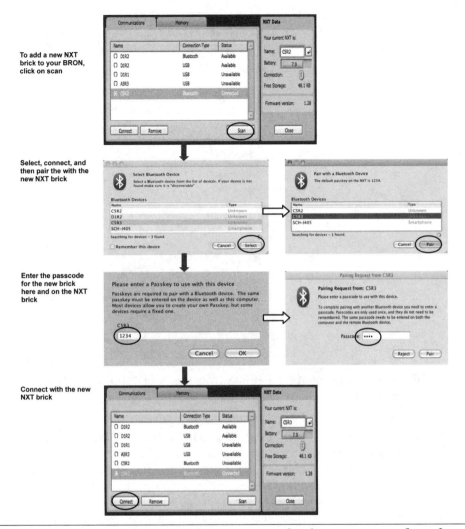

FIGURE 5-5   The steps for connecting a new NXT brick to a computer from the NXT-G's "NXT Window."

- Send Message
- Receive Message

These blocks will allow the NXT bricks to turn on/off, connect/disconnect, send, and receive messages from the team leader or initiator.

NXT-G Connection Block

The NXT-G Connection block turns the Bluetooth communication ability of an NXT brick on or off. Turning the Bluetooth capability on and off can be done from the

FIGURE 5-6 Listing of NXT bricks from the NXT terminal and the status of a selected brick.

NXT brick submenus. But this would mean that the NXT brick would have to be on throughout the program's execution, even when Bluetooth communication was not in use. Bluetooth communication uses the batteries of the brick, causing a drain on battery power. Enabling and disabling Bluetooth communication from within the program can save battery power.

The Connection block offers four options listed in the "Action" section of the configuration panel:

- Turn on
- Turn off
- Initiate connection
- Close connection

The "Turn on" option will turn on the Bluetooth functionality of the brick, allowing it to communicate with another brick (the initiator) with which it has identified and has exchanged passwords. The "Turn off" option disables communication, preventing the brick from communicating with any connected brick or future connections to NXT bricks. Figure 5-7a shows the "On/Off" options of the Connection block.

BLUETOOTH CONNECTION BLOCKS

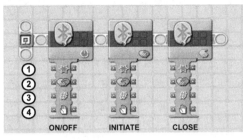

① ACTION

② CONNECT TO

③ CONNECTION #

④ DISCONNECT FROM

CONFIGURATION PANELS

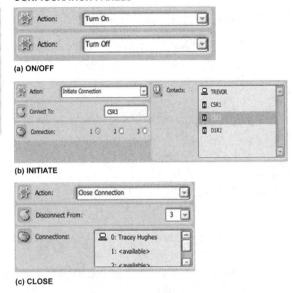

FIGURE 5-7   NXT-G Bluetooth connections: (*a*) "Turn on/off"; (*b*) "Initiate"; (*c*) "Close" options in the "Action" drop-down menu and their configuration panels.

The Bluetooth "Turn on/off" connection option blocks should appear before the first and after the last Send and Receive message blocks.

The "Initiate Connection" option tells the program the names of the bricks with which to communicate (Figure 5-7*b*). The brick can only communicate with other microcontrollers with which it has paired. The brick that performs this initiation procedure will be considered the team leader or initiator. All the names of the available NXT bricks (the computer has made available) will be listed in the "Contacts" list in the configuration panel. The NXT microcontroller can be chosen from the "Contacts" list or typed into the "Connect To" edit box. The brick name is to be associated with a connection or communication port. Remember, all this information should be consistent with the initial setup and pairing performed on the NXT microcontroller initiator. Since the initiator is always associated with line 0, line 0 is not an option to associate with other bricks. The NXT-G software will not prohibit including an Initiation block in the program of a team member. The "Contacts" list will list the names of microcontrollers with which it has identified and exchanged passwords.

The "Close Connection" option will disconnect the Bluetooth connection for the specified brick chosen in the configuration panel (Figure 5-7*c*). A brick can only have a connection with the initiator. The "Connections" list in the configuration panel will list all the names of bricks with which the initiator is currently connected. Lines 0 to 3 can be disconnected. The initiator can close line 0, which will prevent other bricks from communicating with it.

LabVIEW On/Off and Connection Bluetooth Blocks

LabVIEW has separate blocks for turning the Bluetooth connection on and off, connecting and disconnecting Bluetooth. The Bluetooth On/Off block supports these inputs:

- NXT Connection (1)
- On/Off (2)

and these outputs:

- NXT Connection (3)
- NXT BT Power (4)

Figure 5-8a shows the LabVIEW Bluetooth On/Off block.

On/Off Connection Block's Inputs and Outputs

NXT connection input for the On/Off block (1) connects to a previous block in order to establish or continue the program flow. The On/Off input (2) turns the Bluetooth connection of the NXT brick on or off. It is a logical value: **TRUE** turns the communication on, and **FALSE** turns it off.

NXT connection output for the On/Off block (3) connects the current block to the next block. The NXT BT Power Output (4) determines the strength of the Bluetooth communication signal. If the value is 0, the communication signal if off; if it is 1, then the communication signal is on.

Establishing a Connection to the BRON Using LabVIEW

The Connection block connects two NXT bricks via Bluetooth communication. This block support these inputs:

- NXT Connection (1)
- Partner Name (2)
- Connection (5)

and these outputs:

- NXT Connection (3)
- Success (4)

Figure 5-8b shows the Bluetooth Connection block.

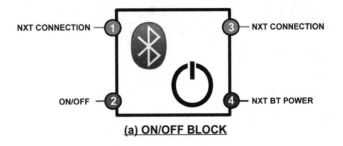

(a) ON/OFF BLOCK

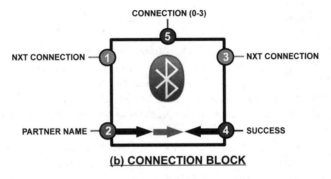

(b) CONNECTION BLOCK

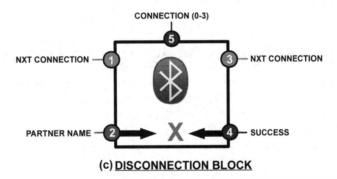

(c) DISCONNECTION BLOCK

FIGURE 5-8 (*a*) The LabVIEW Bluetooth On/Off block, (*b*) Connection block, and (*c*) Disconnection block.

The NXT connection input for the Connection block (1) connects to a previous block in order to establish or continue the program flow. Connection (5) is the communication port or line the NXT bricks uses for its Bluetooth communication. The communication port can be lines 0 to 3. Partner Name is the name of the NXT brick with which the current brick is paired over the communication port.

The NXT connection output for the Connection block (3) connects to the next block in order to establish or continue the program flow. Success (4) is a Boolean value reflecting the result of the connection: **TRUE** if the connection was successful and **FALSE** if the connection failed.

The Disconnection block is a separate block that disconnects two NXT bricks via Bluetooth communication. This block supports these inputs:

- NXT Connection (1)
- Partner Name (2)
- Connection (5)

and these outputs:

- NXT Connection (3)
- Success (4)

Figure 5-8*c* shows Bluetooth Disconnection block.

The NXT connection input for the Disconnection block (1) connects to a previous VI block in order to establish or continue the program flow. Connection (5) is the communication port or line the NXT bricks uses for its Bluetooth communication. The communication port can be lines 0 to 3. Partner Name is the name of the NXT brick with which the current brick is paired over the communication port.

The NXT connection output for the Disconnection block (3) connects to the next VI block in order to establish or continue the program flow. Success (4) is a Boolean value reflecting the result of the disconnection: **TRUE** if the disconnection was successful and **FALSE** if the disconnection failed.

Communicating a Message to the BRON

Once the team leader's brick and the NXT program have been set up for Bluetooth communication, BRON Bluetooth messaging can begin. The team leader can send commands or signals to team members to execute programs, perform maneuvers, activate sensors, and control motors. Members can return messages to the team leader that contain acknowledgments or even data to be sent to other team members.

Sending/Receiving Messages in NXT-G

In NXT-G, the Send Message block allows the brick to send a message to an NXT brick. The "Message" drop-down menu in the configuration panel allows for selection of the type of data sent in a message:

- Number
- Text
- Logic

Figure 5-9 shows the NXT-G Bluetooth Send Message block and the configuration panel for (a) number, (b) text, and (c) logic data types. Once these values are set, they are displayed on the Message block, as shown in Figure 5-9.

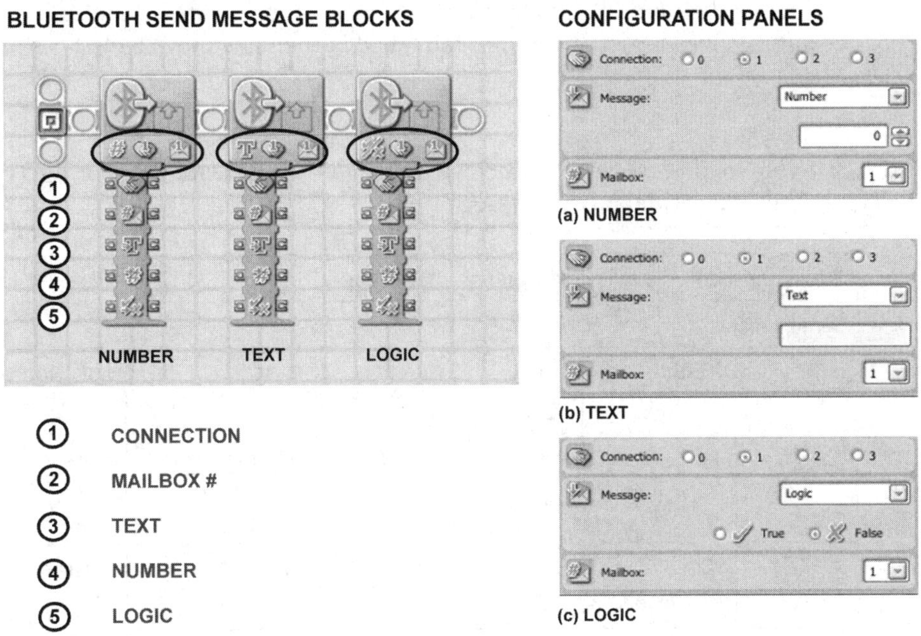

BLUETOOTH SEND MESSAGE BLOCKS

CONFIGURATION PANELS

NUMBER   TEXT   LOGIC

(1) CONNECTION

(2) MAILBOX #

(3) TEXT

(4) NUMBER

(5) LOGIC

(a) NUMBER

(b) TEXT

(c) LOGIC

FIGURE 5-9   NXT-G Bluetooth Send Message block and the configuration panel for (*a*) number, (*b*) text, and (*c*) logic data types.

The "Message" drop-down menu in the configuration panel allows for selection of the type of data that will be sent as a message to the initiator or team leader. Select "Number" if the message to be sent is an integer value, and then enter in the number in the edit box. Select "Text" if the message to be sent is a text message, and then enter the actual text message in the edit box. If the message to be sent is a Boolean value, select "Logic," and then select "True" or "False." Select the mailbox into which the message is to be placed and then received. If the brick is the initiator, it can send messages to other bricks or team members on lines 1 to 3. If the brick is a team member, then messages can only be sent to the initiator on line 0.

Each brick has 10 mailboxes numbered 1 through 10 (Mailbox 1, Mailbox 2, etc.). Up to five values of the same type can be stored in a single mailbox. An attempt to place a sixth value in a mailbox will cause the first (oldest) value to drop off.

Dynamically Setting Values for the Send Message Block

The connection number, mailbox number, and message can be set in the configuration panel. These values also can be set dynamically using the data hub by employing the input data wires. Other blocks can pass these values to the Message blocks by connecting their output data wires to the block's appropriate input data plugs. The incoming values have to be within the range of that data plug. Table 5-1 shows the data plugs, their data types, and ranges.

TABLE 5-1 The Send Message and Receive Message Blocks' Data Plugs, Plug Types, Data Types, and Ranges and a Brief Description

Plug Icon	Data Plug	Plug Type	Data Type	Range	Description
Send/Receive					
	Mailbox	Input/ Output	Number	1–10	Location of stored message
	Text	Input/ Output	String	58 characters maximum length	Message type is a string.
	Number	Input/ Output	Integer	–2147483648 +2147483648	Message type is an integer.
	Logic	Input/ Output	Boolean	True/False	Message type is a Boolean.
Send Data Plugs					
	Connection	Input/ Output	Integer	0–3	Communication port where message is sent to
Receive Data Plugs					
	Message Received	Output	Boolean	True/False	True if message was received, false otherwise
	Comparision	Output	Boolean	True/False	Compare received message to an expected message
	Text Received	Output	String	58 characters maximum length	Message received is a string.
	Number Received	Output	Integer	–2147483648 +2147483648	Message type is an integer.
	Logic Received	Output	Boolean	True/False	Message type is a Boolean.

The Bluetooth Receive Message block allows the brick to receive messages of a specified type from a mailbox. The mailbox has to be specified, but not the communication port. Robots can only receive messages from line 0. If messages are sent to the initiator or team leader, communication port or line 0 is used. If messages are received by the team members, messages can only be sent by the initiator. The initiator may receive a message from any of the team members. Thus, if the initiator must know the name of the sender, the ID of the sender can be a part of the message or can be sent as a separate message in that mailbox or some other mailbox. When using the approach of separate mailboxes for the ID and message, there may be problems with matching the ID with the right message when used by multiple bricks.

The received message can be compared with the contents of the edit box beneath the "Data Type" drop-down menu in the configuration panel, whether it is a number or text. If the received message is a Boolean value, it can be compared with the checked logic value. The result of this comparison can be checked by accessing the "True/False" data plug beneath the "Bluetooth Message Received" data plug on the data hub. Figure 5-10 shows the Bluetooth Receive Message block and configuration panel for number, text, and logic data types.

As with the Send Message block, specified values are also displayed on the block. In this case, though, only the message data type and mailbox number are displayed. The connection number will always be 0.

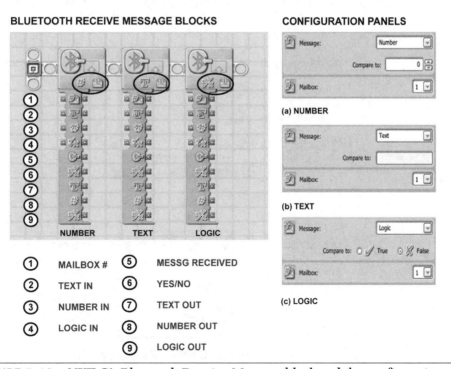

FIGURE 5-10   NXT-G's Bluetooth Receive Message block and the configuration panel for (*a*) number, (*b*) text, and (*c*) logic data types.

Data Hub for Receive Message Block

The Receive Message block's mailbox number and data type can be set dynamically through their input data plugs, but the Receive Message block also has five "output only" data plugs pertaining to the received message. These data plugs are listed in Table 5-1.

Writing/Reading a Message Using LabVIEW

The Write Message block in LabVIEW writes a Bluetooth message to a communication port. This block supports

- Sequence flow in/out
- Connection number
- Mailbox number
- Communicated text
- Status

Figure 5-11*a* shows the Bluetooth Write block.

The Sequence Flow In/Out (1,5) establishes or continues the program flow by connecting with the previous or next block. The connection number and mailboxes

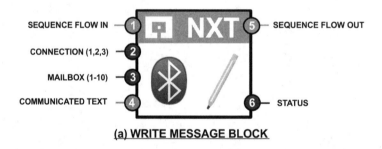

(a) WRITE MESSAGE BLOCK

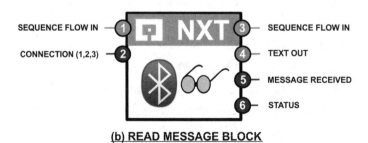

(b) READ MESSAGE BLOCK

FIGURE 5-11 The LabVIEW Bluetooth (*a*) Write and (*b*) Read blocks.

work in the same way as in NXT-G. The connection number (2) represents the communication port to which the NXT brick is connected. The connection number is associated with the NXT brick in which the current brick will be communicating. For the initiator/team leader (host) to send a message to a team member (client), the connection number is from 1 to 3. The Mailbox Number (3) represents the location to which the message is written. There are 10 mailboxes, numbered 1 through 10, associated with each connection number. The Communicated Text (4) is the text that is sent. This text can be 58 characters in length and can contain numbers and characters. The message is converted to a string or flattened and placed in the specified mailbox. The Status (4) is a Boolean value reflecting the result of the sent message: **TRUE** if the message was successfully sent and **FALSE** if the message was not sent.

The Read Message block reads a Bluetooth message from a specified mailbox. This block supports

- Sequence flow in/out
- Mailbox number
- Text out
- Message received
- Status

As in NXT-G, messages received by team members are communicated over Connection or line 0 (team leader/initiator), and therefore there is no need to specify a connection number or communication port. Messages received by the team leader can be sent from any of the team members; therefore, connections 1 through 3 must be specified. Figure 5-11*b* shows the LabVIEW Bluetooth Read Message block.

Text Out (5) is the message that was received by the NXT brick. This message can be 58 characters in length and can contain numbers and characters. Message Received will contain a **TRUE** Boolean value if the message will be in Text Out, **FALSE** if the message is not in Text Out. Status (6) is a Boolean value reflecting the result of the received message.

We've looked at LabVIEW, G, NXT-G, NXC, and JAVA and closely at the coordination and communication with Bluetooth. Once we have all these things in place, how do we put it all together? We'll get a peek at a possible approach in Chapter 6.

Chapter 6

Robot Environments, Teamwork Strategies, and Goals

The Lost Scrolls of Robotics: #6
Now that I have fulfilled my purpose, I don't know what
to do.

—Sonny, *I Robot*

The robots we build don't exist in a vacuum (well, unless we loan them to NASA or one of the other space agencies). For the rest of us, our robots will operate in a particular physical reality. They will function in a particular environment. They will move on a certain type of terrain. They will operate in certain temperatures. The robots might operate indoors or outside in the elements or both. The robots will have either a stationary base or a mobile base. If the base is stationary, then the robot most likely will have actuators and end effectors that move specific distances. Those distances could be measured in millimeters to meters. The robot may operate in a well or poorly lit environment or in sunlight. We call the physical environment where the robot operates the *robot's environment/world*.

The Robot's World

When it comes to environments and building robots, there is often the chicken or the egg question: Do we build the robot based on the environment in which it is going to operate, or do we build the robot as we like and then find a suitable environment? The world in which the robot operates will determine how fast or how slow its actuators can move. It will determine whether sensors work or not. For example, sunlight, ambient room light, and candlelight all affect MINDSTORMS color/light sensors differently, as we discussed in Chapter 3. It's one thing for a robot to move across a wooden floor, and it's entirely another matter for the robot to travel across gravel, dirt, or carpet. Surfaces will affect wheel rotation calculations or distance calculations (Chapter 4). The heights of objects within the physical space will dictate whether the robot sensors or actuators/end effectors will be able to interact with the objects. If the objects are too high, then the ultrasonic or color/light sensors may not be able to see them. When we first started building robots, we built the robots we liked and then tried to make them work in the environments we had available. These days we know well in advance of building a robot what environment the robot will be operating in. Or, if possible, we build an environment to fit the robots we have. Next to the actual robot, the robot's world is the most important consideration. This is especially true in programming teams of robots. Once we entertain the idea of a team of robots, we have to consider whether all team members are compatible with the environment. If the fundamental unit of measure in the environment is a meter and we have some robots that are designed to move meters and some robots that are designed to move centimeters, then there will be more than a few conversion considerations. Robots may have actuators and end effectors that are designed to function at different height levels referred to as the Z axis. The X, Y, and Z axes give us the three-dimensional (3D) space within which the actuators and end effectors have to operate. Depending on how diverse the team is, 3D spatial and distance conversions may be required for communications and messaging between robots.

The robot's world is where robot teams will have to function. Since not all robots are created equal, the success of the team as a whole is tied to each team member being able to navigate within and interact with the world successfully. Thus we cannot talk about building teams of robots without first considering the robots' world where they will operate. Even if our robots are all-terrain (i.e., they can navigate any kind of surface), we must consider whether the objects they will encounter, like our beaker in Chapter 4, will fit the end effectors they have. We also must consider whether the weights of the objects are beyond what the servos/motors can handle. From a distance perspective, it may turn out that the objects are too high or too low and therefore outside the sensor range. The area might be too large! Bluetooth signals are limited to approximately 10 m. If we have an environment that is 10 m wide by 12 m long, and we have robots in each corner of the environment, will they all be able to communicate? Sometimes a very simple model of the environment can help us to get started. Let's take for example the simplest of all robot worlds. Figure 6-1 is a photo of the two-dimensional (2D) robot world that comes with the MINDSTORMS NXT robot kit.

The robot's world is a test pad that is approximately 24 inches wide and 30 inches long. It is rectangular in shape. There are 16 colors on the test pad. There are 38 unique numbers, but some of the numbers are duplicated. There is a series of straight lines and arcs on the pad. During the initial testing of many MINDSTORMS NXT robots, this test pad is the robot's world/environment. If we set the robot's start position on one corner of the test pad and program it to search for a blue square, the robot has no more than 720 in^2 to search depending on which corner is set as the robot's starting point.

The blue square either is in the robot's environment or it's not. Since we stated that this was a 2D environment, the robot only has to search along its X axis or Y axis for the blue square. We assume a value of 0 for the Z axis. Notice that we did not mention

FIGURE 6-1   The 2D MINDSTORMS NXT robot world.

anything about the robot's capabilities. We only gave a physical description of the robot's environment and the task to be executed. The robot designer will have to come up with a design that will meet this challenge. Thus, from this we see that the environment and the problem alone can dictate many aspects of the robot's design. Clearly, in this case, the robot needs to be mobile and have some type of visual sensing—a camera, a light sensor, infrared (IR) sensors, and so on. In addition, the 720 inches says something about the size of the robot that we are building. A robot that has a base of 1,500 in^2 would immediately violate the environmental constraints. Thus the fact of the matter is that if we don't have a detailed model or view of the robot's environment and nature of the problem, then it will be very hard for us to build the robots needed. In fact, the robot's target environment and the nature of the tasks that it will be expected to perform have to be, if not the first question answered, then certainly among the top two or three.

The Robot READ Set

Recall from Chapter 1 the capability matrix that captures the capabilities of all the potential team members. We use the capability matrix as an inventory for all the device capabilities that we have at our disposal. The capability matrix is a kind of visual chart that gives us an indication of the kinds of projects that our robots can handle versus the kinds of projects that they definitely will not be able to handle. In addition to the capability matrix, every robot has to deal with a *robot environmental attribute description* (READ) set, whether the robot builder realizes it or not. The robot READ set is simply the list of objects that the robot will encounter and interact with within the robot's environment. The environment itself has attributes, and each object within the environment will have a certain set of attributes. Let's take for example the test pad that comes with the MINDSTORMS NXT robot kits. In this case, the test pad is the READ set. It consists of several objects, and each object has a set of attribute/ descriptions. Table 6-1 contains the READ set for the MINDSTORMS NXT test pad.

TABLE 6-1   The READ Set for the MINDSTORMS NXT Test Pad

Attribute	Value
Object: Physical Workspace	
Width	24 inches
Length	30 inches
Height	0
Shape	Rectangular
Surface	Paper (smooth)
Object: Color (light)	
Number of colors	16
Light intensities	16
Colors	Red, green, blue, yellow, orange, white, black, gray, green, light blue, silver, etc.
Object: Symbols	
Symbol	Integers
Integer values	0–30, 40, 60, 70, 90, 120, 180, 270, 360
Geometry	Lines, arcs, squares

These three objects and their attributes represent the READ set of whatever the robot will be commissioned to solve when the environment is the test bed. Take for example the problem of identifying and locating a blue square (if there is one). The test bed's READ set represents all the challenges the robot will encounter. The READ set and the robot's capability matrix are two basic tools that will help you to decide:

- If you have the appropriate end effectors, sensors, actuators, and torque and power supplies
- Whether you can even build a robot team to solve the problem

- If you can build the team of robots, the physicality the robots should have collectively

The more automated the robot team is, the more detail of the robot's environment will be required to build the proper robots. If the robots are going to be remote controlled, then the human being controlling the robots can use his or her own vision to navigate the environment and decide which obstacles to move, lift, drag, or carry. The human controller can use his or her own vision to identify shapes and colors, and he or she can use his or her own judgment to determine whether an object is too heavy for the robot to attempt to lift. This can reduce the number and types of sensors the robots will need. This can and will reduce the complexity of the programming needed for each robot. Human controllers reduce the need to have a thorough preknowledge of the environment because human remote controllers can make impromptu assessments and decisions. On the other hand, a team of robots that are fully automated and self-sufficient must know everything about the environment. The more sensors, the better. The more information the robots have, the more likely it is that they will be able to accomplish their goal. The READ set becomes absolutely mandatory.

The READ set captures all the objects in the environment that the robots will interact with. But just listing the objects in the environment does not give us the complete picture of what we're up against. Not all environments are created equal. Some environments are dynamic, changing at the very time the robots are trying to accomplish a task. Objects within the environments are sometimes affected by forces other than the robots. Some environments are either partially, fully, or not accessible to a robot. If all the objects in the environment are within the reach of the robot's sensors and actuators, then the environment is said to be *fully accessible*. If the robot's sensors or actuators can interact with only some things in the environment, then the environment is partially accessible. For instance, an environment may contain sounds and objects. The robots may be able to interact with the objects but have no sensors that can detect the sounds. In this case, the environment is partially accessible. The environment may be deterministic. That is, the environment may be accessible, and the only time it changes is when the robot changes it. There are no outside forces acting on any of the objects. Whatever the state the robot puts the environment in, that's the state the environment will stay in until the robot makes another change. Because not all environments are created equal, we must be concerned not only with the objects in the environment but also with the nature of the environment itself. Table 6-2 contains the basic classifications of environments.

Each class of environment will have its own READ set. For example, the READ set of a dynamic environment will have to include time/location pairs for any objects that may be acted on by outside forces. Consider that a robot's task may be to retrieve a red canister. However, an agent outside the robot may be allowed to change the location of the red canister. This makes the environment dynamic. The robot cannot just simply go to where it expects the red canister to be. The dynamic environment requires the robot to consider when and where it last encountered the canister and

TABLE 6-2 Classifications of Environments

Environment Type	Environment Description
Fully accessible	All aspects of the environment are accessible through the robot's sensors, actuators, or end effectors.
Partially accessible	There are some objects that are not reachable or cannot be sensed, whereas others can.
Deterministic	The next state of the environment is completely determined by the current state and the actions performed by the robot(s).
Nondeterministic	The next state of the environment is not completely under the robot's control; nondeterministic environments have objects that may be influenced by outside factors, or external agents may be added or remove objects from the robot's environment; the environment may be partially accessible, and the robot may not be able to account for objects that it cannot sense or reach.
Episodic	Episodic environments are divided up into self-contained parts; what the robot does in one part does not affect the other sections of the environment; the robot only has to consider the objects and events that pertain to the section that it is in; each section is considered a separate episode.
Static versus dynamic	The environment does not change while the robot is in between actions or if no outside forces act on the environment while the robot is performing an action; if it does change in either circumstance, then the environment is dynamic.
Discrete	The number of actions that the robot must take and the number of objects that the robot must interact with are fixed.

whether it is there now. Put another way, every environment is a little different, and each environment has its own set of challenges. It's been our experience that you cannot realistically separate building and designing your robots from a thorough understanding of their environment. The environment in which the team works will determine how team members can or cannot work together. The fact that there are several types of environments means that any planning of team coordination has to be deferred until the class of the environment and READ set are understood. Putting together a team of robots to accomplish a given goal will be absolutely dictated by the environment classification, the robot's READ set, and the capability matrix.

Robot Application Architecture

The team of robots, the environment, and the task to be performed by the robots make up the *robot application architecture* (RAA). Figure 6-2 shows the major components of the RAA.

Notice in the figure that the RAA has two environments. The first environment is the original environment. This represents the environment before any robots have interacted with it in any way. The second environment is the environment that the robots have interacted with or manipulated in some way. If the robots were successful in executing their task, then the second environment represents the solved problem or completed work. If the team of robots failed, then the second environment can be used to see what went wrong. Was the physical design of the robots appropriate? Did the event programming fail in some way? Were the robot communications improperly designed? And so forth. The RAA represents the complete approach (including solution or set of tasks) that will be necessary to pronounce that the robot teamwork was successful. We have been laying the groundwork in previous chapters for a team-based RAA using MINDSTORMS NXT/Tetrix robotic kits. Throughout the remainder of this book we will use team-based RAA to explain how to build your own team of robots using MINDSTORMS NXT robots and Bluetooth communications.

A Simple Team-Based RAA Example

We would like to know which of the three sodas, Pepsi, Coke, and 7Up, has the most acidity per 20 fluid ounces. Recall from Chapter 2 that D1R2 has a Vernier probe and is capable of measuring the pH of an aqueous substance, but it does not have the

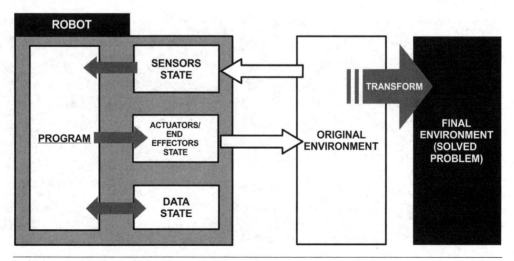

FIGURE 6-2   The major components of the RAA.

capability of retrieving that substance. It can measure and analyze but not retrieve. D1R1 potentially has the capability to identify cans, bottles, or glasses of Pepsi, Coke, and 7Up and might be able to actually transport them to D1R2. We're not sure that D1R1 is up to the task, so we'll look a little closer at the capability matrix. While D1R2 can measure the pH balance, it doesn't necessarily have the onboard processing power to make the comparisons among three different substances. Again, we will have to consult the capability matrix. D3C1 does have the processing power to analyze the results of D1R2, so we will use D3C1 to report which of the three sodas has the most acidity. It is important to note that none of these devices is capable on its own with its current configurations to perform the task we require. The task requires teamwork. One robot identifies and retrieves the substance, another robot analyzes the substance, and we use a computer to evaluate the results of the substance analysis. For this example we will look at the five components that make up the RAA and are used to build a team of robots:

- Capability matrix
- READ set
- Robot designs
- Logic flowcharts
- Robot programs

These five elements give us a complete picture of the team-based RAA for our problem. Once you have a grasp of these five elements, you can begin to build your own team of robots to accomplish whatever challenges you think they're up to!

Figure 6-3 is a close-up of the LEGO temperature probe and Vernier pH sensor probe. The pH probe is 12 cm long. Notice that the Vernier probe is considerably larger and heavier. There are NXT-G blocks and LabVIEW blocks that read this sensor. Although there is not a specific leJOS class for them, the raw values can be read and

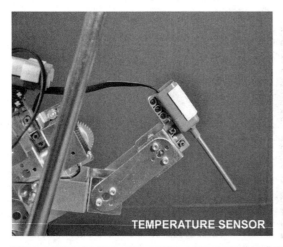
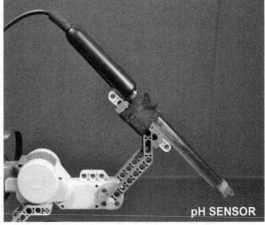

TEMPERATURE SENSOR

pH SENSOR

FIGURE 6-3 The LEGO temperature probe and Venier pH sensor probe.

a simple formula can be applied to those values to get reasonable voltages. We will cover in detail later how the leJOS `SensorPort` classes are used by setting the type to `RAW_MODE` and reading `rawValues()`. For now, Table 6-3 will just remind you of some of the pH sensor's specifications from Chapter 3.

TABLE 6-3   Some of the pH Sensor's Specifications

Specification	Value/Range	Description
Response time	90 percent of final reading in 1 second	The amount of time it takes the sensor to return a reading as output to the user
Temperature range	5–80°C	The operating temperature range in which readings are reliable
Range	pH 0–14	Range of the reading output
Isopotential pH	pH 7	Point at which temperature has no effect
Output	59.2 mV/pH at 25°C	Corresponding the pH to the millivolts; with a 7 pH corresponding to 0 mV, each pH change corresponds to a change of ±59.2 mV.

We constructed an arm and holder for the pH sensor that is attached to D1R2. When comparing the arm and holder with the other robots' arms and end effectors, it was evident that they would not be able to hold and manipulate the pH probe. Figure 6-4 shows close-ups of the robots' arms and end effectors for D1R1 and A3R3 and the arm and holder for D1R2. D1R1's gripper could not hold the probe, and A3R3's gripper could not support the length of the probe, especially when the probe was not in immediate use. A3R3 simply would lower the probe too fast to the surface whenever it was powered down or when the arm was changing positions. We didn't want A3R3 to damage the tip of the probe by hitting the surface too hard. D1R2's holder also could accommodate other similar Venier probes.

The Multipurpose Capability Matrix

We often built the robot first and then decided what neat things we could do with it. In fact, that's precisely how the vast majority of our robots have been built. Shoot first; ask questions later. As the sophistication of our robots grew, though, and the cost of parts started to go up, we had to be a little more concerned with what we had to work with prior to actually building the robots. The capability matrix answers the following questions:

- Can the robot and parts I have do the job?
- In order to do the job, what sensors, parts, actuators, and so on are missing?

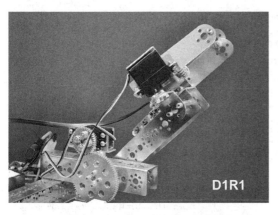

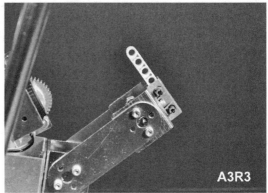

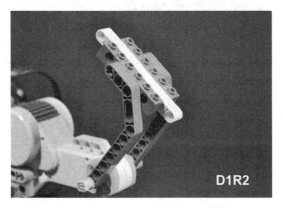

FIGURE 6-4 The arms and end effectors of D1R1 and A3R3 and the arm and holder of D1R2.

The capability matrix also generates at least one more question:

- With these robots and parts, what kinds of things can I do?

The answer to this question is where most robot hobbyists/builders start off. No matter which of the three questions you look at, the major use of the capability matrix falls under the category of planning. Any good robot application architecture has a planning component. Table 6-4 shows the capability matrix for our pop testing application.

TABLE 6-4 The Capability Matrix for Our Pop Testing Application

Team D (Robots/Devices)	Microcontroller/Microprocessor	End Effectors	Mobility type	Display	Sensors/Motors/Controllers	I/O Port	Ranges/Degrees/Capacity	Communication
D3C3	AMD Turion 64-bit dual 2.8-GHz processor	N/A	Portable	19-in HD liquid-crystal display (LCD)	N/A	N/A	1.5 terabytes of storage	802/11b wireless Bluetooth USB 10/100 Ethernet modem, infrared
D1R1	32-bit ARM7, 64-kB RAM	Arm and gripper	Four wheels	100- × 64-pixel LCD screen	LEGO touch	3	1–0	Bluetooth, USB
					HiTechnic color	2	16 colors	
					LEGO ultrasonic	4	0–255 cm with a margin of ±3 cm	
					Direct-current (DC) motor controller	1	360 degrees	
					Left wheel DC motor	DC controller 1–1+	360 degrees	
					Right wheel DC motor	DC controller 2– 2+	360 degrees	
					Left wheel encoder	ENC. 1	360 degrees	
					Right wheel encoder	ENC. 2	360 degrees	
					Arm servo	Servo controller port 1	90 degrees	
					Gripper servo	Servo controller port 2	90 degrees	

(continued on next page)

TABLE 6-4 The Capability Matrix for Our Pop Testing Application (*continued*)

Team D (Robots/ Devices)	Microcontroller/ Microprocessor	End Effectors	Mobility type	Display	Sensors/ Motors/ Controllers	I/O Port	Ranges/Degrees/ Capacity	Communication
D1R2	32-bit ARM7, 64-kB RAM	Probe and touch sensor extension	Two tractor wheels	100- × 64- pixel LCD screen	LEGO touch	1	1–0	Bluetooth, USB
					pH sensor adapter	3	pH 1–14	
					Probe servo	B	90 degrees	
					Right tractor wheel servo	C	360 degrees	
					Left tractor wheel servo	A	360 degrees	

A Basic READ Set for D1R1, D1R2, and D3C1

We actually have to look at the capability matrix and the READ set at the same time to decide which robots will be able to contribute and in what ways they will be able to contribute to the team. Recall that the READ set will include the physical environment within which the team of robots will have to operate. Figure 6-5 is the top view of a simple READ set for our pop analysis task.

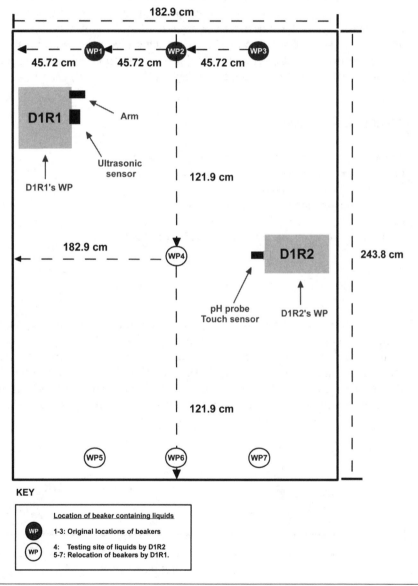

FIGURE 6-5 Top view of the READ set for the pop analysis task.

Some of the important features of the READ set are that it is 5 × 6 ft in size. It's a rectangular area. There will be three 9-cm-tall glass beakers with a diameter of $^7/_{12}$ cm. The beakers are going to be placed 2 ft apart on the left edge of the environment. We can see where the robots will be placed in the environment. Remember, the READ set is also a major planning tool. Figure 6-6 is a side view of the READ set. The side view is important because it gives you the relative heights and lengths of all the objects that the robots have to deal with in proportion to the heights of the robots.

Looking at the READ set top and side views and the basic goal of the robots, we decide which robots we will use for the project. Once that decision is made, we then have to come up with a plan or strategy that actually will result in the robots accomplishing the task at hand. There are several common methods for developing a rough draft for what the robot should do. For example, when using pseudocode, we might have statements like the following:

PSEUDOCODE EXAMPLE

```
Robot D1R1 waits for message from computer.
Once robot has message it proceeds to waypoint
If robot is at a waypoint then
    pick up object
else
    keep traveling on current course heading
end if
.
.
.
```

We call these partially English, partially computer-type language lists *pseudocode*. Pseudocode allows you to specify a narrative of what the robot is supposed to do without having to use a program language and without worrying about a particular language syntax. In our case, we chose flowcharts because we think they are easier to follow. To borrow from the old saying, a picture is worth, well, at least 10 lines of code! The flowchart allows us to visually specify what we want our team of devices to do. Figure 6-7 shows the simplified flowchart for D1R1's tasks.

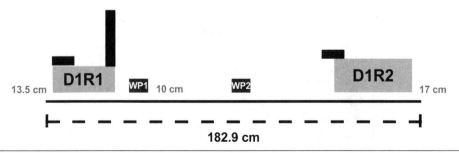

FIGURE 6-6 Side view of the READ set for the pop analysis task.

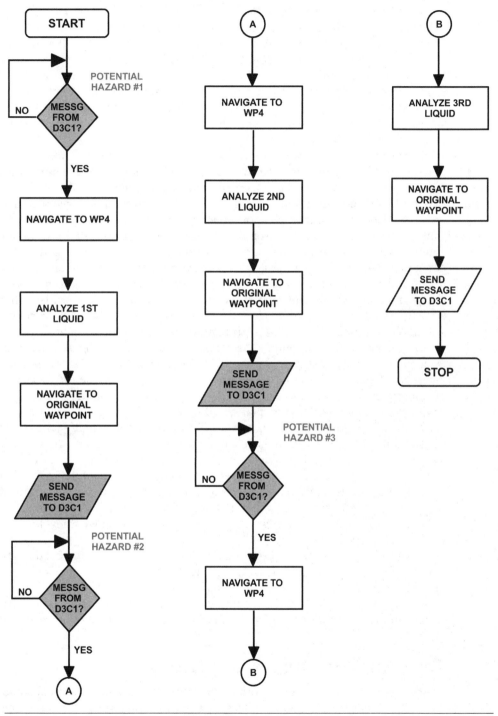

FIGURE 6-7   Flowchart for D1R1's tasks.

Recall D1R1's job is to locate three containers of pop and take them one by one to D1R2 to be analyzed. We start at the start symbol and proceed until we hit the stop symbol. The flowchart is a rough draft of what the robot should do. The question is, Does the flowchart really represent the major tasks that D1R1 has to do? For example, how will D1R1 communicate with D1R2 that the beaker is ready to be analyzed, especially since D1R1 and D1R2 are not directly connected and do not send data directly back and forth? Does the flowchart reflect the communication? Figure 6-8 shows the flowchart for D1R2's tasks. This gives us a visual look at what we think the main tasks are for D1R2.

Notice that D1R2 in Figure 6-8 and D1R1 in Figure 6-7 do not make a move until they receive a Bluetooth message from D3C1. Thus, if for some reason the messaging fails, no work gets done. Although we all know that the team is only as strong as its weakest link, communication failures, or failing to plan for a message that takes a little extra time, or failing to plan for the situation in which the wrong message is sent, continues to be the primary reason why team robot applications fail. Also notice in Figure 6-8 that there are a number of hazard spots for D1R2. The hazard spots show potential weaknesses in our design plan. They are potential hazard spots. If we put in code to cause an infinite wait to eventually time out and if we put in verifications for the message reading and writing, we will avoid the potential hazards. Can you see what the problem might be in Figure 6-8? Therein lies the value of flowcharts—they are supposed to show you potential design problems prior to actually writing code.

On this team, the computer's primary role is to compare the analysis of the three pops and simply to relay messages between D1R1 and D1R2, as shown in Figure 6-9. Since D1R1 and D1R2 are not directly connected, the only way they can communicate with each other is through D3C1. Thus that communication should be the center of attention when planning for error prevention or error detection. If we look at the flowcharts in Figures 6-7, 6-8, and 6-9 and visually follow what we think should be happening, we have some idea whether our plan will work without having to write a single line of code. In this case, the level of detail in the flowcharts is moderate. We could have drilled down even deeper. For example, in the flowchart in Figure 6-7, where D1R1 returns to its original location, we could have further specified that the next step is to do a 180-degree turn so that it would be facing the right direction for the next waypoint. It's up to you to decide what level of detail you'll need in your flowcharts.

Another important piece of information that we could have included in D1R1's flowchart is the speed of the robot. Each container holds a liquid, and we don't want the liquid to splash, so we set the motor speed to an acceptable value. This kind of information might (or might not) be important to determining whether your final design will work. But it's good to give consideration to this level of information.

Teamwork Strategies and Goals

Before digging deeper into teamwork strategies, let's take a closer look at some of the environment classifications. In this book, we are concerned with fully accessible environments. We make sure that the environments are fully accessible because they

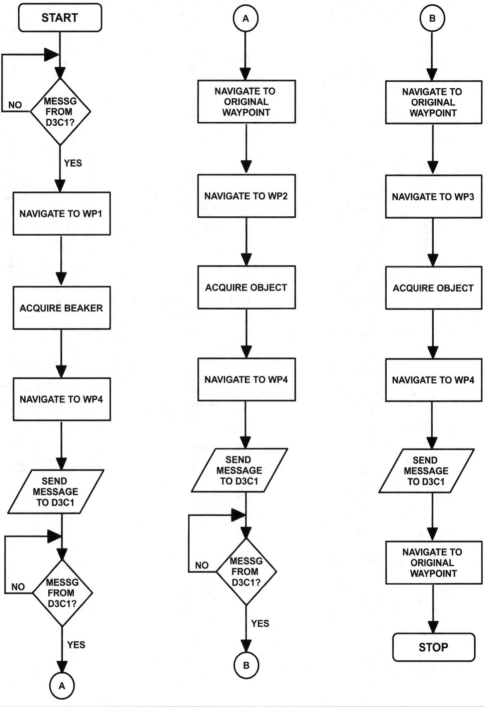

FIGURE 6-8   The flowchart D1R2's tasks.

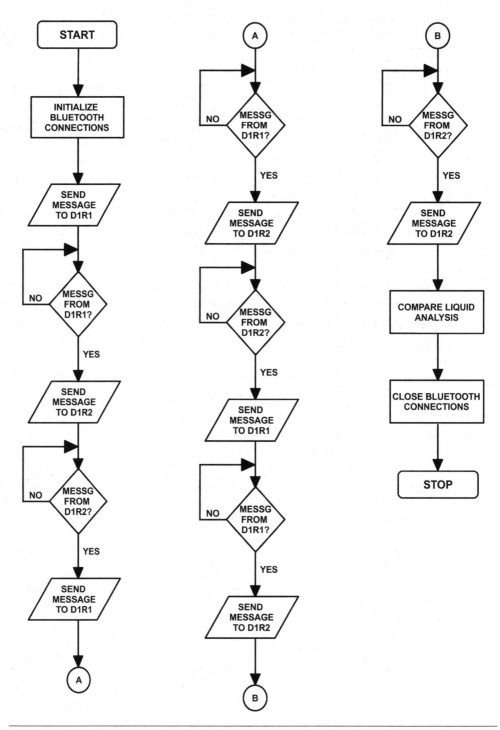

FIGURE 6-9 The flowchart for D3C1.

are built specifically for our robots to operate in. The teamwork strategies and goals are absolutely dictated by the READ set and the type of environment in which the team will operate. The teams that we demonstrate in this book are fully autonomous. There are no remote controls. Once the robot starts, it will continue on its own until it either finishes its program or there is some kind of failure. Ideally, we want the environment to be fully accessible, deterministic, and episodic. If we can predict what the environment will look like at any given time based only on the robot's current, previous, and future actions, then we have a deterministic environment. The more autonomous you want your robot to be, the more accessible and deterministic you will want the robot's environment. However, like environments, autonomy comes in many flavors. Table 6-5 describes the levels of autonomy.

TABLE 6-5   Autonomy Levels

Autonomy Level	Autonomy Type
1	Simple rule-based
2	READ set (environment) + robot program
3	Simple goal-based + robot program
4	READ set (environment) + goal + robot program

Simple Rule-Based Autonomy and READ Set + Robot Program Autonomy

If a robot has simple rule-based autonomy, the robot considers only its current (immediate) situation, and then the robot executes whatever rule(s) apply to that situation. The immediate situation includes

- What is sensed right now
- What the current location is
- What the motors are doing at this instant

And based on this information, the robot has been programmed to take a specific action. In simple rule-based autonomy, the robot does not concern itself with actions it has already taken. The robot does not concern itself with how the environment was or what the environment will be at some point in the future. The robot concerns itself only with the here and now and the sensors, actuators, and end effectors that are immediately relevant.

Note   When just learning how to program a team of robots, we strongly recommend that you give all robots the same kind of autonomy. Mixing and matching autonomy types can lead to dynamic, nondeterministic, partially accessible environments (or to put it succinctly—trouble).

If a robot has READ set + rule-based autonomy, the robot considers the environment it is in, including any changes that the robot may make to the environment in deciding which rule it should activate or which action to take next. In the READ set + rule-based autonomy, the robot keeps track of how it has changed the environment and what the current environment looks like based on changes the robot has made. The current environment dictates what actions the robot will take. Figure 6-10 contrasts the simple rule-based autonomy with READ set + rule-based autonomy.

SIMPLE RULE-BASED AUTONOMY

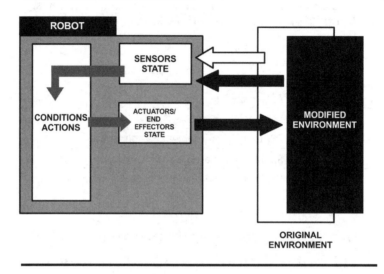

READ SET + RULE-BASED AUTONOMY

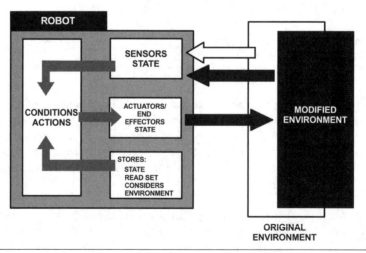

FIGURE 6-10 Simple rule-based autonomy versus READ set + rule-based autonomy.

Notice in Figure 6-10 that in simple rule-based autonomy, the robot does not keep track of any changes that it has made to the environment. It does not store the READ set or anything about the environment. It simply responds to the program based on current sensor readings. For example, the robot may be programmed to move forward until the touch sensor hits an object. If the touch sensor hits an object, then the robot turns around and moves forward until the touch sensor hits another object. In this example, the robot does not store where it has been, where it is, or even whether it hit an object or not. It just responds to the program. The robot may hit the same object over and over again until it runs out of power, or if it is in a large enough space, it may continue without hitting an object until it runs out of power, falls off an edge, and so on. If it is asked to move an object from location X to location Y, it will not remember where the object is if you ask it to later move the object to location Z. In fact, it won't remember what the object was that it moved. On the other hand, the other RAA in Figure 6-10 does remember the location of the object and what object it moved. Program + READ set autonomy stores the current values of all the objects in its READ set, and those values will be used to help the robot carry out its next action. For example, if we ask a robot with level 2 autonomy to move a red ball from location X to location Y, and then we ask it to move the red ball to location Z, a level 2 autonomous robot will remember that it is currently at location Y. A level 1 autonomous robot will have to be directed where to find the ball. It will not remember that it moved it from location X to location Y. The levels of autonomy in Table 6-5 are progressive. A robot with level 4 autonomy is more autonomous than a robot with level 1 autonomy. The program for a level 2 robot will require more memory because details about the original environment and the current environment are stored in the robot's random access memory (RAM). These details are used by the robot's program to execute the task.

Level 3 and Level 4 Autonomous Robots

A robot with level 3 autonomy is aware of its READ set. It typically has a *single* goal. It remembers the result of its actions on the environment, but instead of having a predetermined sequence of actions to take (e.g., step 1, then step 2, then step 3, etc.), a level 3 autonomous robot has a list of legitimate actions that it may take without any particular order necessarily specified. The robot decides which action to take based on its goal and its current environment. Whichever action moves the robot closer to its goal is the action that the robot will take next. The level 4 autonomous robot typically has two or more goals, and sometimes the goals may even be conflicting. The level 4 autonomous robot is aware of its READ set—it remembers its actions. Further, it plans its actions in order to make the goals easier to achieve. Based on the environment, its potential impact on the environment, and its available actions, the level 4 robot will select which goal or goals to execute and then which actions are necessary to achieve those goals. In level 1 and level 2 autonomy, the robot's next action is already predetermined simply based on the state of the environment. The robot cannot decide which action to take; the environment really dictates what the next action is. In level 3 and level 4 autonomy, the robot decides the next action* based on how favorable that action plus the current state of the environment is to the robot's goal(s). In this book, we only build robots that have level 1 and level 2 autonomy. Designing and building level 3 and level 4 autonomous robots is not for the faint of heart and is beyond the scope of this book. The simplest approach in programming the autonomous team is to make every robot on the team either level 1 or level 2. Although this is not absolutely necessary, it makes the team programming easier.

Environment, READ Sets, and the Team Challenge

Recall from Chapter 1 that we introduced three basic types of coordination between team members:

- Time-based or chronology-based coordination
- Event-based coordination
- Message-based coordination

Now that we have a clearer picture of the part that the environment and the READ set play in building and programming our robots, we can see what a critical part these types of coordination play. Because the robots all are intersecting with the same environment, each has the potential to change that environment. If Robot A has a complete picture of what the environment is, and then Robot B changes that environment, how will Robot A's picture of the environment be updated? What if there are three robots on the team? If Robot B changes the environment, how will Robot A's and Robot C's pictures of the environment be updated? Here's where our Bluetooth communications come in. In order for all the team members to be on the same page when dealing with the environment, any changes that are made by any of the robots must somehow be made known to the other robots on the team. While this is less of a problem for a level 1 autonomous robot because it typically is not programmed based on the state of the environment, it is an absolute necessary for level 2 autonomy and higher. With the basic configurations that we have used throughout this book, there are two approaches to keeping the environmental/READ set information current for all team members. Figure 6-11 shows the first approach.

In this figure, the computer is used as the team leader. We take advantage of the computer's memory capacity to store the changes to the environment for each robot and to keep a consistent picture of the environment for each robot. Each robot gets

---

*The decision of which action to take next is a major area of research in robotics and agent-based programming. It covers such topics as swarm intelligence, evolutionary computing, genetic algorithms, and biologically inspired behaviors. These topics are far beyond the scope of this book. However, if the reader is a glutton for punishment, see *Bio-Inspired Artificial Intelligence*, by Dario Floreano and Claudio Mattiussi, for a somewhat kind and merciful introduction to these topics.

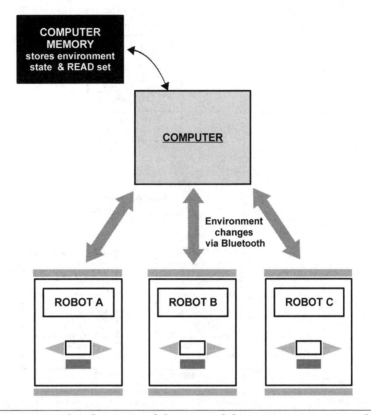

FIGURE 6-11   Centralized storage of the state of the environment using the computer as the team leader.

the environment information it needs from the computer and sends any changes that it makes to the environment to the computer. The computer then updates the state of the environment. In this configuration, there is only one state, and the READ set is stored on the computer, and Bluetooth messaging is used between the computer (team leader) and the NXT bricks (team members) to keep a shared picture of the environment. The second basic approach is shown in Figure 6-12.

In this second approach, each team member has its own copy of the state of the environment. As each robot makes changes to the environment, it not only updates its own state, but it also sends the new values to the team leader. The team leader then is responsible for sending the new values to each team member. Each team member updates its own copy of the environment. In this way, all team members will have the same view of the environment as they make changes. In this case, Bluetooth communications are absolutely necessary. Otherwise, Robot A might attempt to acquire the red object from the specified location and be totally unaware that Robot B has already moved the red object from the specified location. Although there are other techniques that we could use to ensure that all team members maintain the same

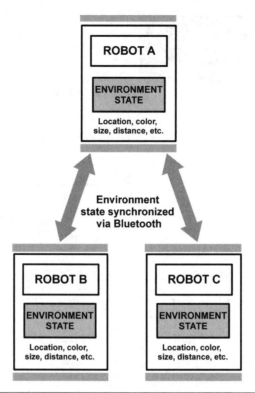

FIGURE 6-12 Each robot has its own copy of the state of the environment synchronized using Bluetooth communications.

view of the environment, the approaches shown in Figures 6-11 and 6-12 are the two used most often and act as a good foundation.

Let's Not Fool Ourselves, It's Slow!

Bluetooth is known for its simplicity and convenience, not necessarily its speed. The more information you transfer between robots, the longer it takes. In some cases it feels like the robots have come to a complete stop, when in actuality they are communicating environment changes. So there is some tradeoff: The more autonomy we want the robots to have, the more environmental information is needed and updated. The more complex the tasks the team of robots needs to perform, the more communication is necessary. In our case, therefore, the speed at which the robots communicate is somewhat of an occupational hazard that we have to live with for now or at least until the Bluetooth standard is optimized in some new way.

A Closer Look at a Level 2 Autonomous MINDSTORMS/Tetrix-Based Team

We're going to look at D1R1 and A3R3. Figure 6-13 is a photo of D1R1 and A3R3. In the figure, A3R3 has a LEGO temperature probe, a HiTechnic compass, and an ultrasonic sensor, and D1R1 has a HiTechnic color sensor, a robot arm with gripper end effector, and an ultrasonic sensor. Both D1R1 and A3R3 have Tetrix chassis, servos, and DC motors with encoders.

FIGURE 6-13   Photos of D1R1 and A3R3.

The basic processing in maintaining the current view of the state is to update the state every time the robot makes a change in the environment or a change occurs in the robot's state. Here D1R1 and A3R3 will both maintain their own copies of the state of the environment. They will be using techniques from Figure 6-3. But instead of three NXT bricks, the team will consist of a computer (D1C3) as team leader and two NXT bricks as team members. Recall that in this configuration, D1C3 (the computer) is the initiator, and the two NXT bricks are the receivers. Each brick will update its own state and picture of the environment and then send those new values to D1C3. D1C3 then will send the values to the other teammate. D1C3 also will keep its own copy of the environment and the changes that were made by D1R1 and A3R3.

How Do We Know When the Task Is Done?

The team leader keeps track of the robots' progress toward executing the task. Since the team leader will always know the current READ set and the environment, as well as the current state for all the team members, the team leader will always be in a position to know when the task is done or whether the team has failed and it's time to communicate the need to give up.

If you take a closer look at our pop testing project, you'll see that it is written in leJOS and that all the robot programming is divided into three major classes. Each Java class represents a team member. The team leader in this case is a computer. We chose the computer as the team leader because it brings the processing power to the team to do the substance analysis that we need performed. Each class represents a team member, and each class represents a "software version" of the robot. Each class has a software component for every hardware component that the robot has. Listing 6-1 shows the Java declaration for each team member:

LISTING 6-1

```
 1   import java.io.DataInputStream;
 2   import java.io.DataOutputStream;
 3   import java.io.BufferedReader;
 4   import java.io.InputStreamReader;
 5   import lejos.pc.comm.NXTCommLogListener;
 6   import lejos.pc.comm.NXTConnector;
 7   import lejos.pc.comm.NXTInfo;
 8   import lejos.pc.comm.NXTComm;
 9   import lejos.pc.comm.NXTCommFactory;
10   import lejos.pc.comm.NXTCommBluecove;
11
12
13   // Team Leader
14
15   public class d3c1
```

```
16  {
17     String D1R1 = "D1R1";
18     String A3R3 = "A3R3";
19     String D1R2 = "D1R2";
20     float Temperature[];
21     DataOutputStream Dout;
22     DataInputStream Din;
23     int OneSecond = 1000;
24     NXTComm D1R1Comm;
25     NXTComm A3R3Comm;
26     NXTInfo D1R1Robot;
27     NXTInfo A3R3Robot;
28
29     ...
30
31
32
33
34     public d3c1() throws Exception
35     {
36        Temperature = new float[3];
37        D1R1Comm = NXTCommFactory.createNXTComm(NXTCommFactory.
                   BLUETOOTH);
38        A3R3Comm = NXTCommFactory.createNXTComm(NXTCommFactory.
                   BLUETOOTH);
39     //NXTComm  D1R2Comm = NXTCommFactory.
                          createNXTComm(NXTCommFactory.BLUETOOTH);
40        if(D1R1Comm == null){
41
42           System.out.println("Could Not find Brick");
43           System.exit(1);
44
45        }
46        else{
47
48              System.out.println("Found Brick" + D1R1);
49
50        }
51        D1R1Robot = new NXTInfo(NXTCommFactory.BLUETOOTH,D1R1,
                   "00:16:53:0C:54:91");
52        A3R3Robot =  new  NXTInfo(NXTCommFactory.BLUETOOTH,A3R3,
                   "00:16:53:01:A9:8D");
53
54     }
55
```

```
56
57
58   //Team Member that has the robot Arm that will transport
59   //the Liquid
60
61
62   public class d1r1_p1 implements NavigationListener
63   {
64       public UltrasonicSensor    Vision;
65       public ColorHTSensor       ColorVision;
66       int CurrentColor;
67       double  WheelDiameter;
68       double TrackWidth;
69       DifferentialPilot  D1R1Pilot;
70       ArcMoveController  D1R1ArcPilot;
71       TetrixControllerFactory  CF;
72       TetrixMotorController MC;
73       TetrixServoController SC;
74       TetrixRegulatedMotor LeftMotor;
75       TetrixRegulatedMotor RightMotor;
76       TetrixServo  Arm;
77       TetrixServo  Gripper;
78       OdometryPoseProvider Odometer;
79       Navigator D1R1Navigator;
80       boolean PathReady = false;
81       Pose CurrPos;
82       int OneSecond = 1000;
83       Sound  AudibleStatus;
84       BTConnection Btc;
85       DataInputStream dis;
86       DataOutputStream Dout;
87
88
89       public d1r1_p1() throws InterruptedException
90       {
91          Vision = new UltrasonicSensor(SensorPort.S3);
92          ColorVision = new ColorHTSensor(SensorPort.S2);
93          LCD.drawString("Sensors  constructed",0,0);
94
95
96          WheelDiameter = 7.0;
97          TrackWidth = 34.5;
98
99          CF = new TetrixControllerFactory(SensorPort.S1);
```

```
100         LCD.drawString("Tetrix Controller Factor
                         Constructed",0,0);
101
102         MC = CF.newMotorController();
103         SC = CF.newServoController();
104         LeftMotor = MC.getRegulatedMotor(TetrixMotorController.
                       MOTOR_1);
105         RightMotor = MC.getRegulatedMotor(TetrixMotorController.
                       MOTOR_2);
106         LeftMotor.setReverse(true);
107         RightMotor.setReverse(false);
108         LCD.drawString("motors Constructed",0,0);
109
110
111
112
113         Gripper = SC.getServo(TetrixServoController.SERVO_2);
114         Arm = SC.getServo(TetrixServoController.SERVO_1);
115         LCD.drawString("Servos Constructed",0,0);
116         Thread.sleep(3000);
117
118         SC.setStepTime(7);
119         Arm.setRange(750,2250,180);
120         Arm.setAngle(100);
121         Thread.sleep(3000);
122
123         Gripper.setRange(750,2250,180);
124         Gripper.setAngle(50);
125         Thread.sleep(1000);
126         D1R1Pilot = new DifferentialPilot(WheelDiameter,TrackWidth,
                                  LeftMotor,RightMotor);
127         D1R1Pilot.reset();
128         D1R1Pilot.setTravelSpeed(6);
129         D1R1Pilot.setRotateSpeed(30);
130         D1R1Pilot.setMinRadius(4);
131
132         LCD.drawString("Pilot Constructed",0,0);
133         Thread.sleep(3000);
134         CurrPos = new Pose();
135         CurrPos.setLocation(0,0);
136         Odometer = new OdometryPoseProvider(D1R1Pilot);
137         Odometer.setPose(CurrPos);
138         D1R1Navigator = new  Navigator(D1R1Pilot,Odometer);
139         D1R1Navigator.addNavigationListener(this);
```

```
140          D1R1Navigator.singleStep(true);
141          LCD.drawString("Odometer Constructed",0,0);
142          Thread.sleep(5000);
143
144
145      }
146
147
148
149      ...
150
151
152
153  }
154
155
156  //  Team Member that will do some of the analysis of the substance
157
158  public class a3r3_p1
159  {
160      public UltrasonicSensor    Vision;
161      public CompassHTSensor     Compass;
162      public TemperatureSensor   SubstanceTemperature;
163      int CurrentColor;
164      double  WheelDiameter;
165      double TrackWidth;
166      DifferentialPilot  a3r3Pilot;
167      TetrixControllerFactory  CF;
168      TetrixMotorController MC;
169      TetrixServoController SC;
170      TetrixRegulatedMotor LeftMotor;
171      TetrixRegulatedMotor RightMotor;
172      TetrixServo  Arm;
173      OdometryPoseProvider Odometer;
174      Navigator a3r3Navigator;
175      Pose CurrPos;
176      int OneSecond = 1000;
177      Sound  AudibleStatus;
178      float  DegreesSubstance[];
179      BTConnection Btc;
180      DataInputStream dis;
181      DataOutputStream Dout;
182
183
```

```
184
185
186        public a3r3_p1() throws InterruptedException
187        {
188            Vision = new UltrasonicSensor(SensorPort.S4);
189            Compass = new CompassHTSensor(SensorPort.S3);
190            SubstanceTemperature = new TemperatureSensor(SensorPort.S2);
191            LCD.drawString("Sensors constructed",0,0);
192            DegreesSubstance = new float[3];
193            WheelDiameter = 7.0;
194            TrackWidth = 34.5;
195            CF = new TetrixControllerFactory(SensorPort.S1);
196            LCD.drawString("Tetrix Controller Factor Constructed",0,0);
197            MC = CF.newMotorController();
198            SC = CF.newServoController();
199            LeftMotor = MC.getRegulatedMotor(TetrixMotorController.
                        MOTOR_1);
200            RightMotor = MC.getRegulatedMotor(TetrixMotorController.
                        MOTOR_2);
201            LeftMotor.setReverse(true);
202            RightMotor.setReverse(false);
203            LCD.drawString("motors Constructed",0,0);
204            Arm = SC.getServo(TetrixServoController.SERVO_2);
205            LCD.drawString("Servos Constructed",0,0);
206            Thread.sleep(3000);
207
208            SC.setStepTime(7);
209            Arm.setRange(750,2250,180);
210            Arm.setpulseWidth(1500);
211            Thread.sleep(3000);
212            Arm.setAngle(100);
213            Thread.sleep(3000);
214
215            a3r3Pilot = new DifferentialPilot(WheelDiameter,TrackWidth,
                                              LeftMotor,RightMotor);
216            a3r3Pilot.setTravelSpeed(6);
217            a3r3Pilot.setRotateSpeed(30);
218            LCD.drawString("Pilot Constructed",0,0);
219            Thread.sleep(3000);
220            CurrPos = new Pose();
221            CurrPos.setLocation(0,0);
222            Odometer = new OdometryPoseProvider(a3r3Pilot);
223            Odometer.setPose(CurrPos);
224            a3r3Navigator = new Navigator(a3r3Pilot,Odometer);
```

```
225        a3r3Navigator.singleStep(true);
226        LCD.drawString("Odometer Constructed",0,0);
227        Thread.sleep(5000);
228
229
230    }
231
232    ...
233
234
235
236
237
238 }
```

If we look at lines 62 through 87, we see the major components of D1R1, the team member that is responsible for transporting the substance. This is the typical approach used throughout this book in associating the robot's software capabilities with the robot's hardware capabilities. Another basic technique that we deploy is to use the class **Constructor** to start up all the sensors and motors. The class **Constructor** is also used to set up the initial state of the robot's READ set variables. For example, to set up the robotic arm, lines 209 and 210 state

```
209        Arm.setRange(750,2250,180);
210        Arm.setpulseWidth(1500);
```

These two lines act as a calibration for the robot arm. They set the range of the arm and the rough midpoint of that range. Although this can be changed or set at any time, it is usually good practice to set this kind of stuff in the class **Constructor**. To set the state of the READ set, use statements such as

```
220        CurrPos = new Pose();
221        CurrPos.setLocation(0,0);
222        Odometer = new OdometryPoseProvider(a3r3Pilot);
223        Odometer.setPose(CurrPos);
```

that set the initial position of one of the team members. Also notice the import statements in lines 6 through 10. They show the NXT communication headers that are necessary to use Bluetooth communications between NXT bricks and the computer. Look at lines 34 through 54. These statements are part of the constructor for D3C1 (the team leader), and these are the statements that are used to set up Bluetooth communications to the two NXT bricks, D1R1 and A3R3. Examples of the most critical lines here include

```
A3R3Comm = NXTCommFactory.createNXTComm(NXTCommFactory.BLUETOOTH);
A3R3Robot =  new  NXTInfo(NXTCommFactory.BLUETOOTH,
                       A3R3,"00:16:53:01:A9:8D");
```

where `00:16:53:01:A9:8D` is the address of the NXT brick named `A3R3`.

Listing 6-2 shows some of the more interesting methods of team member A3R3.

LISTING 6-2

```
1   public void waitUntilStop(int Distance) throws Exception
2   {
3
4       Distance = Math.abs(Distance);
5       Double TravelUnit = new Double(Distance/a3r3Pilot.
                             getTravelSpeed());
6       Thread.sleep(Math.round(TravelUnit.doubleValue())
                         * OneSecond + 100);
7       a3r3Pilot.stop();
8
9   }
10
11  public void waitForRotate(double Degrees) throws Exception
12  {
13
14      Degrees = Math.abs(Degrees);
15      Double DegreeUnit = new Double(Degrees/a3r3Pilot.
                             getRotateSpeed());
16      Thread.sleep(Math.round(DegreeUnit.doubleValue())
                         * OneSecond + 100);
17      a3r3Pilot.stop();
18
19
20  }
21
22
23
24  void analyzeSubstance(int Liquid)  throws Exception
25  {
26      travel(-5);
27      waitUntilStop(-5);
28      moveArm(0);
29      moveArm(80);
30      Thread.sleep(7000);
31      DegreesSubstance[Liquid] = getDegrees();
```

```
32      moveArm(0);
33      travel(5);
34      waitUntilStop(5);
35
36  }
37
38
39
40  void  reportSubstanceAnalysis() throws Exception
41  {
42
43      AudibleStatus.twoBeeps();
44      String connected = "Connected";
45      String waiting = "Waiting...";
46      String closing = "Closing...";
47      LCD.drawString(waiting,0,0);
48      LCD.refresh();
49      Btc = Bluetooth.waitForConnection();
50      LCD.clear();
51      LCD.drawString(connected,0,0);
52      LCD.refresh();
53      dis = Btc.openDataInputStream();
54      Dout = Btc.openDataOutputStream();
55      int n = dis.readInt();
56      analyzeSubstance(0);
57      Dout.writeFloat(DegreesSubstance[0]);
58      dis.close();
59      Dout.close();
60      dis = Btc.openDataInputStream();
61      Dout = Btc.openDataOutputStream();
62      dis.readInt();
63      analyzeSubstance(1);
64      Dout.writeFloat(DegreesSubstance[1]);
65      dis.close();
66      Dout.close();
67      dis = Btc.openDataInputStream();
68      Dout = Btc.openDataOutputStream();
69      dis.readInt();
70      analyzeSubstance(2);
71      Dout.writeFloat(DegreesSubstance[2]);
72      //Thread.sleep(OneSecond * 5);
73      dis.close();
74      Dout.close();
75      Btc.close();
76
```

```
77
78
79  }
80
81  public static void main(String [] args)  throws Exception
82  {
83
84
85      a3r3_p1 a3r3Robot = new a3r3_p1();
86      a3r3Robot.reportSubstanceAnalysis();
87
88  }
```

The `reportSubstanceAnalysis()` shows the Java code that is used to wait on a Bluetooth connection from D3C1 and how to receive a message and how to send a message back to the team leader.

In lines 49 through 59, the Java code that waits on a connection, sends a message, writes a message, and analyzes substances is shown:

```
49          Btc = Bluetooth.waitForConnection();
50          LCD.clear();
51          LCD.drawString(connected,0,0);
52          LCD.refresh();
53          dis = Btc.openDataInputStream();
54          Dout = Btc.openDataOutputStream();
55          int n = dis.readInt();
56          analyzeSubstance(0);
57          Dout.writeFloat(DegreesSubstance[0]);
58          dis.close();
59          Dout.close();
```

Listings 6-1 and 6-2 show some of the interesting processing that our robot team performs. The complete source code listing for these three robots can be downloaded from www.robotteams.org. The NXT robots on the team are controlled by D3C1. Here in Listing 6-3, are the main functions that control A3R3 and D1R1:

LISTING 6-3

```
1  public void analyzeLiquid(int CurrentLiquid) throws Exception
2  {
3
4      getMessageD1R1(CurrentLiquid);
5      sendMessageA3R3(CurrentLiquid);
```

```
 6        getMessageA3R3(CurrentLiquid);
 7        sendMessageD1R1(CurrentLiquid);
 8
 9
10
11  }
12
13  public static void main(String[] Args) throws Exception
14  {
15
16        d3c1 Computer = new d3c1();
17        Computer.openConnectionD1R1();
18        Computer.openConnectionA3R3();
19        Computer.analyzeLiquid(1);
20        Computer.analyzeLiquid(2);
21        Computer.analyzeLiquid(3);
22
23
24
25  }
26
```

One of the important notes to make about Bluetooth communications between two or more team members is to remember that you only need to open the Bluetooth connection once. Although you may open and close the input and output streams any number of times during execution of the team program, only open and close the Bluetooth connection between team members once, unless you have reasons to intentionally open and close the connection during the course of the program. Once you get the idea of Bluetooth programming between the NXT bricks and a computer, the same approach is used between the NXT and the Android smart phone or any Linux/Java-based device that has Bluetooth communication capabilities.

Robots Do Only What They Are Told

BRON'S BELIEVE IT OR NOT!

Robots follow their programming. So why do we sometimes get behavior from robots that we don't expect? Or how is it that we program the robot to do one thing, and it does the exact opposite? Sometimes it seems like no matter what we put in the program, the robot just doesn't work. So there has to be something else to it, right? Well, yes and no. There are situations where a hardware failure prevents the robot from properly executing its program, making it look like the robot is acting on its own without your programming. For example, if an encoder is malfunctioning, or if a gear is stripped, or if the power source is almost completely drained, these types of things can interfere with the proper execution of the program. So if the robot is behaving in a manner other than what is expected, make sure that there is no faulty hardware and no hardware failure. Our NXT microcontrollers have several layers of firmware/software that we did not write. There is the NXT-OS firmware or leJOS firmware, NXC software libraries, and leJOS class libraries.

In many cases the robot is following the software with precision. The real problem is that we don't fully understand what an NXC procedure is really doing, or we didn't look closely at the documentation (where it exists) for some leJOS class. One of the primary reasons for getting bad behavior from your robot is traceable to the fact that there is something in a piece of the software that you didn't write and that you really don't understand. So, in addition to ruling out a hardware malfunction, be completely sure that you understand every function, procedure, class method, or library call that you make. Have a detailed understanding of how each piece of code works if your robot depends on it. Finally, make sure that you didn't leave out any program steps. Your robot is not a mind reader (at least not yet). For the most part, it only does what you program it to do, not what you want it to do. Make sure that what you program the robot to do and what you want it to do are exactly the same thing!

We did leave ourselves a little breathing room here. We did say "for the most part, it only does what you program it to do." There are a couple of other forms of programming that are increasingly being applied to robots that include *agent-oriented programming* and *evolutionary programming*. These approaches to programming involve giving the robot goals, objectives, beliefs, desires, and intentions. The robot then uses these to figure out what its next course of action will be. The robot's behavior or set of actions is not driven by specifically programming what the robot is to do and when but rather by what course of action the robot will move closer to a goal or an intention. Agent-oriented and evolutionary programming are among the newest and most promising ways to get robots to execute advanced behaviors (possibly even behaviors that we don't expect but do appreciate). When a robot is executing an agent-oriented program or an evolved algorithm, it most definitely will take proper actions that it was not explicitly programmed to do!

Chapter 7

Give Your Team of Robots Java Power with leJOS

The Lost Scrolls of Robotics: #7

I am not malfunctioning, you are.

—Hector, *Saturn 3*

In addition to Java, in this book we have used NXC, LabVIEW, and NXT-G to power our teams of robots. The important point that we want to demonstrate is that you can build and program your team of robots using any of the available languages. As long as the language supports Bluetooth, you can use most of the techniques presented in this book. NXC, LabVIEW, and NXT-G are all well suited to program the NXT bricks. Thus we suggest using whichever environment you feel most comfortable with, and you should learn as much as you can about it so that you can get the most out of programming your robots. However, if we were asked which of the languages that work with the NXT is the most flexible, is the most expressive, has the most support for talking to other Bluetooth devices, and has the most built-in functionality, then we would say that leJOS Java wins hands down.

Brief History of Java Virtual Machine for MINDSTORMS

leJOS started out as the TinyVM Java virtual machine by Jose Solorzano. TinyVM was an open-source project that later morphed into what is now known as leJOS. Some of the major contributors to leJOS as we now have it were Brian Bagnall, Jurgen Stuber, and Paul Andrews. The leJOS project basically gave us a firmware replacement for

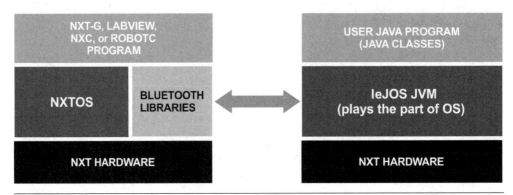

FIGURE 7-1 Basic layering of the leJOS setup for the NXT brick compared with a brick that is set up for NXTOS.

the NXT firmware. It also gave us a rich set of Java class libraries, plug-ins, and code compatible with LEGO, HiTechnic, MindSensors, and Vernier sensors. In addition to these, leJOS gave us several important utilities that make Flashing NXT bricks, uploading and downloading code, and monitoring programs fairly painless. There are two main divisions of the Java libraries that come out of the leJOS project: the `lejos.nxt` libraries, which consist of code that runs on the NXT bricks, and the `lejos.pc` libraries, which consist of code that runs on a computer that is then able to connect to, communicate with, and control the NXT bricks. Figure 7-1 shows the basic layering of the brick when set up for leJOS compared with a brick that is set up for NXTOS.

The Power of leJOS Java for MINDSTORMS NXT

One of the biggest advantages of using the Java environment is the sheer amount of code that is already written for you. While LabVIEW and NXT-G provide lots of ready-to-use blocks and LabVIEW offers some nice sub-virtual instruments (subVIs), they are lacking when it comes to the powerful class libraries that are available in the Java environment. Table 7-1 provides a sample of some of the power functionality that comes built-in with the leJOS environment.

TABLE 7-1 leJOS Built-In Java Classes

leJOS Java Classes	Class Description
Sensor classes	Classes that are designed to interface with, access, and manage sensors. There are classes for LEGOs sensors, HiTechnic sensors, and MindSensors sensors; compass classes, temperature classes, etc.

TABLE 7-1   leJOS Built-In Java Classes (*continued*)

leJOS Java Classes	Class Description
Motor classes	Classes that have full functionality to program motors, set power, set distance, and set speed. There are leJOS classes for the basic NXT motors and the Tetrix direct-current (DC) motors and there are Tetrix encoders as well as Tetrix servos.
Navigator classes	These classes use dead reckoning to keep track of a robot pose (the location in the plane and the robot's heading—the direction in which it moves). Can perform three elementary movements in a plane—travel in a straight line, move in the arc of a circle, and rotate in place.
Steering Pilot classes	Vehicles that are controlled by the `SteeringPilot` class use a steering mechanism similar to a car's, in which the front wheels pivot from side to side to control direction. If you issue a command `travel(1000)` and then issue a command `travel(−500)` before it completes the `travel(1000)` movement, it will call stop, properly inform movement listeners that the forward movement was halted, and then start moving backward 500 units. This makes movements from the `SteeringPilot` leakproof and incorruptible.
GPS classes	These classes interface with global positioning capabilities of the robot and make programming GPS applications easier.
Shortest Path Finder classes	This class calculates the shortest path from a starting point to a finish point while avoiding obstacles that are represented as a set of straight lines. The path passes through the end points of some of these lines, which is where the changes in direction occur. Since the robot is not a point, the lines representing the obstacles should be lengthened so that the actual robot will miss the actual obstacles. It uses modification of the A* algorithm, which is a variant of the Dijkstra shortest path algorithm. This variant adds the nodes needed. It uses the Node inner class for its internal representation of points.
Waypoint classes	A sequence of waypoints makes up a route that a robot can navigate. `Waypoint` extends `Point` because a waypoint can just be a point. However, a waypoint optionally can specify a heading that the robot must achieve when it reaches the waypoints. It also can optionally specify how close the robot must get to the waypoint for it to be deemed to have reached it.
Position classes	Represents the location and heading (direction angle) of a robot. This class includes methods for updating the position in response to basic robot movements. It also contains utility methods for use in navigation, such as the direction and distance to a point from the location of the pose, as well as the location of a point at a given distance and direction from the location of the pose.
Robot Arm classes	A generic interface for robot arms that can move to points in three-dimensional (3D) space. Makes no assumption about the design of the arm.

The classes in Table 7-1 are just a sampling of the classes that come ready to use in the leJOS implementation for MINDSTORMS NXT. The code in these classes literally can save you days, weeks, and months in programming your robots. The availability of these classes and many more is one of the primary reasons why we prefer to use Java for any major robot programming that is necessary.

A Closer Look at the leJOS Utilities

In the main directory of your leJOS installation—in our case it was `/leJOS_NXJ_0.9`
`.1beta`—there will be a subdirectory named `bin` that will contain:

`nxj`	`nxjc`	`nxjconsole`	`nxjcontro`
`nxjdebugtool`	`nxjflashg`	`nxjlink`	`nxjmonitor`
`nxjpcc`	`nxjupload`	`nxjbrowse`	`nxjchartinglogger`
`nxjconsoleviewer`	`nxjdataviewer`	`nxjflash`	`nxjimage`
`nxjmapcommand`	`nxjpc`	`nxjsocketproxy`	

These are all utilities that make up the leJOS environment. And most of them can
be executed individually; for example,

`nxjc hello_world.java`

will compile a java program named `hello_world.java`. The program

`nxj hello_world`

will download the `hello_world` program (without running it) to the brick. The
program

`nxj -r hello_world`

will download and then run the program on the brick. The program

`nxjflash`

will remove the NXTOS from your brick and replace it with leJOS Java virtual
machine (JVM), as in Figure 7-1. Don't worry about the fact that if for some reason
you want to go back to the NXTOS, you can just re-download the firmware to the brick
using the MINDSTORMS NXT-G environment.

Another handy utility is

`nxjmapcommand`

This program is shown in Figure 7-8. This utility is used to help build maps for
your robots to follow and navigate. It is used with the **Navigator**, **Pilot**, and
Pose classes. This is a power utility that allows you to visually design your *robot
environmental attribute description* (READ) set landscape and territory. The commands

`nxjpcc`
`nxjpc`

are used to compile PC programs that can work with the NXT controller and the leJOS
environment. The **nxjpcc** command compiles, and **nxjpc** is used to run the Java
program.

Note If Java directories include the right libraries, the regular `javac` and `java`
commands can be used. But if you don't want the hassle of figuring out all the
classes needed for NXT, including in Bluetooth classes, it simply is easier to use

`nxjpcc` and `nxjpc`. The basic commands/utilities for using leJOS are listed in Table 7-2.

TABLE 7-2   The Basic leJOS Commands/Utilities

leJOS Utilities	Description
`nxjflash`	The command-line utility used to update the NXT controller with leJOS firmware. This effectively replaces the NXTOS with a JVM.
`nxjflashg`	This is the graphical version of `flashg`. We did get it to work with one of our bricks, but in general the command-line version was much more reliable. Once your brick is in update mode, the command-line version works with no problems.
`nxjbrowse`	A graphical tool designed to operate over NXT's file system.
`nxjconsole`	This is the command center for most of the leJOS utility functionality. It servers a similar purpose to the NXT-G command center.
`nxjmonitor`	This is a part schematic, part setup graphical environment used to look at sensors, test motors, etc.
`nxjmapCommand`	Used to graphically build waypoints, landscapes, paths, and maps for NXT robots.

Figures 7-2 through 7-8 contain screenshots of the most commonly used leJOS utilities.

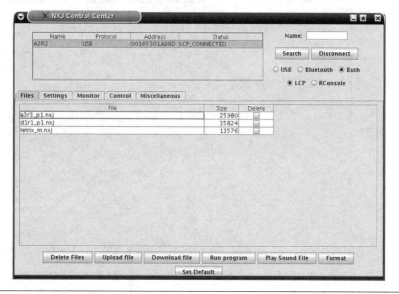

FIGURE 7-2   leJOS utility nxjcontrol: Files.

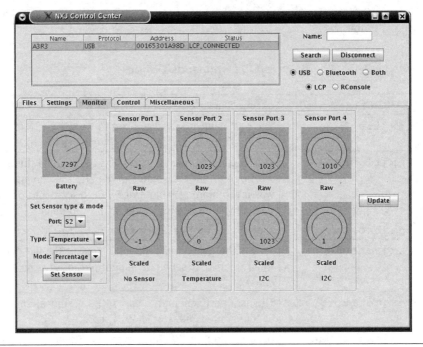

FIGURE 7-3 leJOS utility nxjcontrol: Monitor.

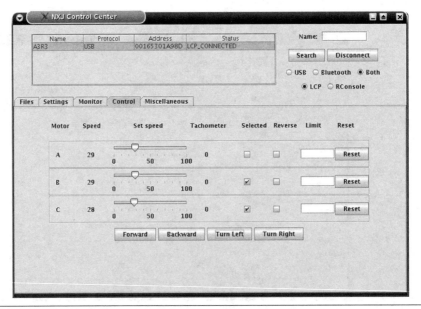

FIGURE 7-4 leJOS utility nxjconsole: Control.

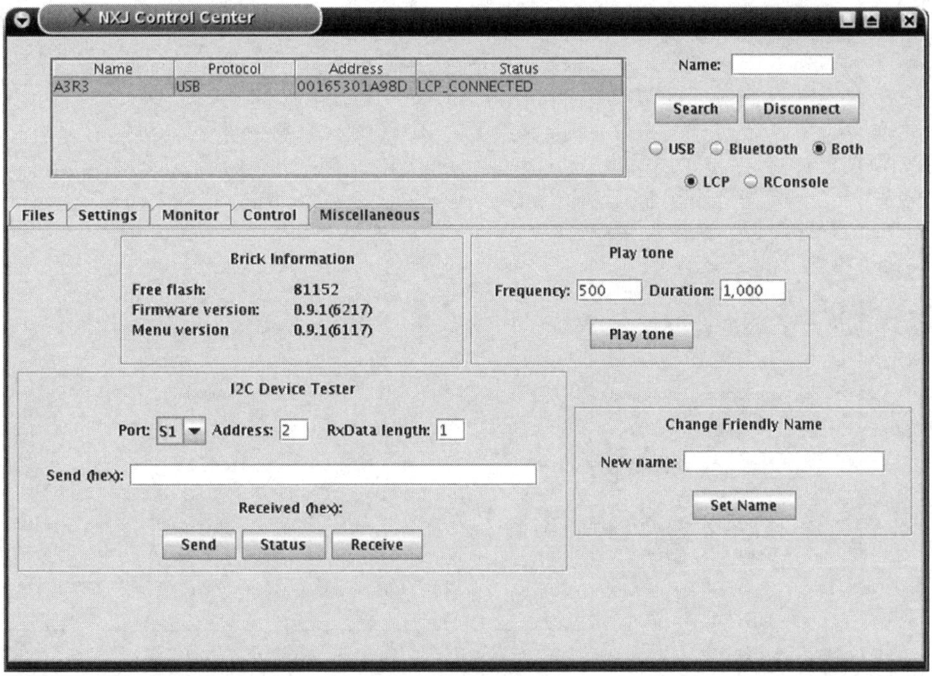

FIGURE 7-5   leJOS utility nxjcontrol: Miscellaneous.

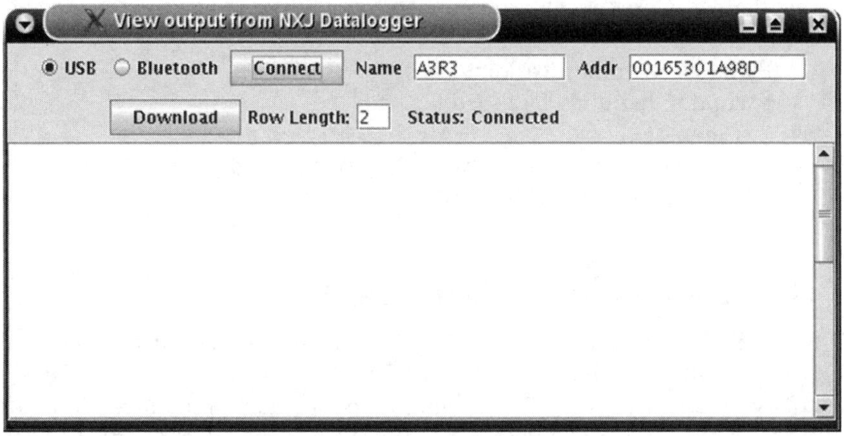

FIGURE 7-6   leJOS utility nxj Datalogger.

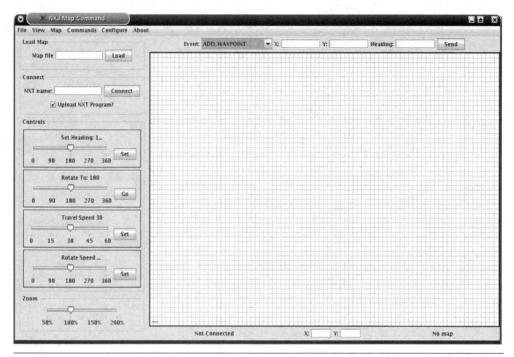

FIGURE 7-8 leJOS utility nxjmapcommand.

Power of Java for Building Teams

The Java language is a big language with a lot to offer. For building our own robot teams based on the MINDSTORMS/Tetrix kits, though, there are five major features that really are essential where Java makes programming a piece of cake:

- Bluetooth communications
- Exception-handling facilities
- Listener facilities
- Inheritance and method overloading (team mode)
- READ set and robot modeling

Bluetooth Communications

The first big advantage to using Java is that Java allows our *Bluetooth Robotic-Oriented Network* (BRON) or piconet to use seven devices instead of three (the standard configuration with NXTOS). In addition to increasing the device number from three to seven, we also can have multiple BRONs (a scatternet) connected. Exactly! We can connect one BRON to another to form more sophisticated, more powerful teams of robots. Another big advantage of using Java is that very few other Bluetooth-enabled devices speak NXT-G or LabVIEW. Java is the de facto language of many small mobile smart devices. For example, the Android smart phone that we use in our Rubik's Cube project runs Linux as its native operating system and speaks Java as its native language. When the NXT brick is running a JVM, it opens the robot up to a completely new world consisting of smart phones, personal digital assistants (PDAs), computers, and so on.

 It is important to note that Bluetooth communications using the leJOS and NXT bricks do not currently use the "Mailbox" concept that is a mainstay of the NXT brick when using NXT-G or LabVIEW. Instead, Bluetooth communications between NXT bricks and other Java-based devices use the concept of Input Streams and Output Streams to communicate. Thus, rather than mailboxes, streams are used. Streams are a powerful concept in Java. They take some getting used to, but once you learn them, you will find them indispensable.

Here is a simple initiator-receiver program that uses leJOS classes and Bluetooth to communicate between two MINDSTORMS/Tetrix-based robots using a computer as the initiator. This is an example of two NXT-based robots that are not in master-slave mode but still need to communicate data to each other. A3R3 sends compass and temperature information to D1R1 by way of the computer (D1C3).

The Robots (A3R3, D1R1)

Figure 7-9 is a side view of A3R3, one of our MINDSTORMS/Tetrix-based robots.

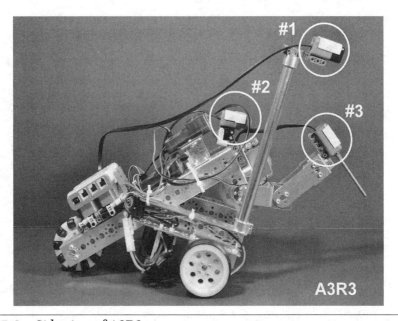

FIGURE 7-9   Side view of A3R3.

Robot (A3R3) In addition to the NXT microcontroller, A3R3's sensor #1 is a HiTechnic compass sensor. Sensor #2 is a stock LEGO ultrasonic sensor, and Sensor #3 is a LEGO MINDSTORMS temperature sensor. A3R3 uses two dc motors from Tetrix, and it has a Tetrix frame and robotic arm. Figure 7-10 shows the stock robotic arm that is part of the Mantix build that we used for A3R3. However, we designed and built a special end effector so that we could attach the LEGO temperature sensor. Figure 7-11 shows the end effector we attached. The end effector was constructed from the Tetrix hard-point connector, the 32-mm channel, and one LEGO 1- × 7-in beam. A3R3 is a perfect example of keeping to your basic design but enhancing the robot so that it can participate in a team. We added the temperature sensor so that A3R3 could participate on the crime scene investigation (CSI) team. A top view of D1R1 is shown in Figure 7-12.

Robot (D1R1) D1R1 has a single NXT microcontroller. Sensor #1 is a HiTechnic barometric sensor, Sensor #2 is a stock LEGO ultrasonic sensor, and Sensor #3 is a HiTechnic color sensor that gives us 16-color capabilities. In addition to the sensors, D1R1 has a robot arm and a gripper for the end effector. D1R1's build is taken from the stock Ranger build from Tetrix. We have, however, replaced the types of sensors used in the original build and redesigned the robot arm, shortening and repositioning it. Notice in the top view of D1R1 that the robotic arm is on the flush-left side of the robot. The stock build has the robot arm in the center of the chassis. This interfered with the ultrasonic sensor's view port. We could have moved the ultrasonic sensor, but being centered served our purpose better, so we chose to move the arm. Thus D1R1 is very similar to the Tetrix-based range finder with a few improvements that make it more compatible with our CSI team.

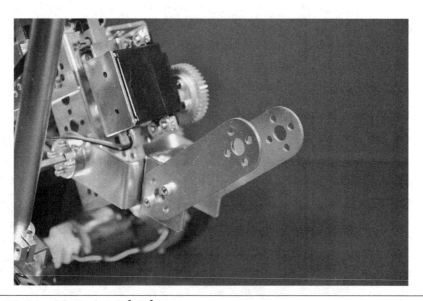

FIGURE 7-10 Mantix's stock robotic arm.

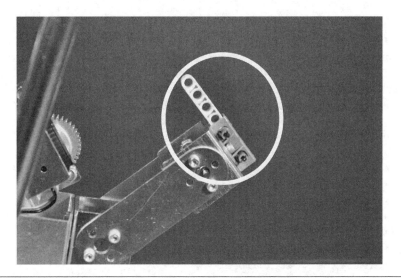

FIGURE 7-11   The end effector for the Mantix robot arm.

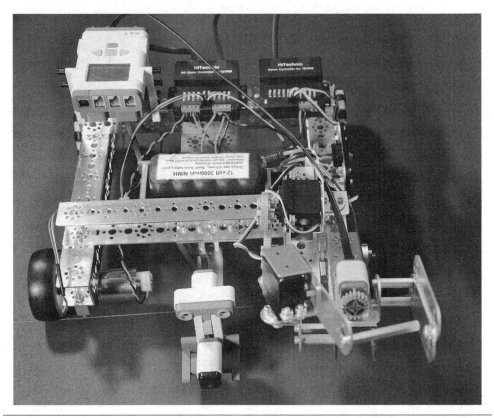

FIGURE 7-12   Top view of D1R1.

The Java Classes

One of the nice features of Java is the notion of class interfaces. Virtually every LEGO/HiTechnic/Vernier sensor or actuator has a Java class. And to access the actuator, sensor, or end effector, you only have to get the class and call the methods (procedures/functions) of that class. For example, A3R3 has a HiTechnic compass sensor, a LEGO temperature sensor and ultrasonic sensor, and Tetrix dc motors and servos. Therefore, in leJOS, we use the following classes:

- `CompassHTSensor`
- `TemperatureSensor`
- `UltrasonicSensor`
- `TetrixMotorController`
- `TetrixEncoderMotor`
- `TetrixServo`
- `TetrixServoController`

Each of these classes will have methods that allow you to manipulate the connected sensor, actuator, or end effector. For example, Table 7-3 lists the methods of the `CompassHTSensor`, `TemperatureSensor`, and `UltrasonicSensor` classes.

TABLE 7-3 The Methods of the Classes `CompassHTSensor`, `TemperatureSensor`, and `UltrasonicSensor`

Method	Description
CompassHTSensor methods	
getDegrees()	Returns the directional heading in degrees.
getDegreesCartesian()	Cartesian coordinate systems increase from 0 to 360 counterclockwise, but compass readings increase clockwise.
resetCartesianZero()	Changes the current direction the compass is facing into the zero angle for the method getDegreesCartesian().
startCalibration()	Starts the calibration for the compass.
stopCalibration()	Ends the calibration sequence.
TemperatureSensor methods	
getResolution()	Returns current resolution.
getSamplingDelay(int resolution)	Returns how long it takes the sensor to measure the temperature at the given resolution.
getTemperature()	Returns the temperature as a floating-point number.
setResolution(int resolution)	Sets current resolution.

TABLE 7-3   The Methods of the Classes `CompassHTSensor`, `TemperatureSensor`, and `UltrasonicSensor` (*continued*)

`UltrasonicSensor` methods	
`capture()`	Sets capture mode; sets the sensor into capture mode.
`continuous()`	Switches to continuous mode.
`getActualMode()`	Returns the current operating mode of the sensor.
`getCalibrationData (byte[] data)`	Returns 3 bytes of calibration data.
`getContinuousInterval()`	Returns the interval used in continuous mode.
`getData(int register, byte[] buf, int off, int len)`	Executes an I2C read transaction and waits for the result.
`getDistance()`	Returns distance to an object.
`getDistances(int[] dist, int off, int len)`	Returns an array of distances.
`getFactoryData(byte[] data)`	Returns 10 bytes of factory calibration data.
`getMode()`	Returns the current operating mode of the sensor.
`off()`	Turns off the sensor.
`getRanges()`	If the sensor is capable, this method returns multiple range values from a single scan.
`getRange()`	Gets the range to the nearest object.
`ping()`	Sends a single ping.
`reset()`	Resets the device.
`sendData(int register, byte[] buf, int off, int len)`	Executes an I2C write transaction.
`setCalibrationData (byte[] data)`	Sets 3 bytes of calibration data.
`setContinuousInterval (int interval)`	Sets the ping interval used when in continuous mode.
`setMode(int mode)`	Sets the sensor into the specified mode.

Once you get accustomed to using one class, you will find it easier to use any class because the interfaces to the sensor and actuator classes are pretty similar. And all the functionality that is related to that particular sensor, actuator, and end-effector is a part of that Java class. So find the right class, and you're in business. To see how all this works, let's look at a simple example. A3R3 will take two temperature readings of two different liquids. It will then send those readings to D1C3, which is the computer. D1C3 then D1C3 will select the highest temperature reading and send that reading and its corresponding heading to D1R1. D1R1 will retrieve the values and then obtain the container holding the liquid.

The Robot Class

The easiest design decision when using leJOS with MINDSTORMS NXT robots is to represent each robot as a Java class. You can either have a generic robot class that every MINDSTORMS robot inherits, or you can make a special class for each robot that you're building. For this simple project, we make a special class. Listing 7-1 contains a partial declaration of our robot class for A3R3.

LISTING 7-1

```
1    import java.io.DataInputStream;
2    import java.io.DataOutputStream;
3    import lejos.nxt.UltrasonicSensor;
4    import lejos.nxt.*;
5    import lejos.nxt.SensorPort;
6    import lejos.nxt.LCD.*;
7    import lejos.nxt.comm.BTConnection;
8    import lejos.nxt.comm.Bluetooth;
9    import lejos.nxt.addon.ColorHTSensor;
10   import lejos.nxt.addon.CompassHTSensor;
11   import lejos.robotics.navigation.DifferentialPilot;
12   import lejos.robotics.localization.OdometryPoseProvider;
13   import lejos.nxt.addon.tetrix.*;
14   import lejos.nxt.addon.tetrix.TetrixRegulatedMotor;
15   import lejos.robotics.navigation.Pose;
16   import lejos.robotics.navigation.Navigator;
17
18   public class a3r3_p1
19   {
20       public UltrasonicSensor    Vision;
21       public CompassHTSensor     Compass;
22       public TemperatureSensor   SubstanceTemperature;
23       int CurrentColor;
24       double  WheelDiameter;
25       double TrackWidth;
26       DifferentialPilot  a3r3Pilot;
27       TetrixControllerFactory  CF;
28       TetrixMotorController MC;
29       TetrixServoController SC;
30       TetrixRegulatedMotor LeftMotor;
31       TetrixRegulatedMotor RightMotor;
32       TetrixServo  Arm;
33       OdometryPoseProvider Odometer;
34       Navigator a3r3Navigator;
```

```
35    // Methods go  here.....
36    }
37
```

Notice that we gave the class the same name as the robot. A3R3 is the robot's name, so in this case we name the class similarly. Also, if you just look at the list of items under the name, you can see what kinds of parts A3R3 has. Right away we see it has an ultrasonic sensor, a compass sensor, and a temperature sensor. If we look further down the list, you can see that A3R3 has Tetrix regulated motors, an arm, and an odometer, and something called a navigator. This is one of the nice features of Java that is not available in NXT-G or LabVIEW, namely, the ability to design your programs as models. The class concept can be used to model all kinds of objects, robots, smart phones, and robot arms, as well as items that will be in the robot's environment, such as maps, paths, routes, and so on. This makes organizing and thinking about your robot's program very straightforward. In our case, we have a robot named A3R3, so we build a class called **a3r3_p1** with the same parts that our robot has. Every hardware component of our robot has a software counterpart in the Java class.

It is important to note the import statements in Listing 7-1. Almost every sensor, actuator, motor, etc. has Java code associated with it. These import statements tell the compiler to include that code with your program. If you do not have all the necessary import statements, your program will not compile.

Initialize the Robot Parts and the READ Set in the Constructors

Remember, in Java, the constructor is used to initialize the attributes of the class and any other important startup functions that are required by the class. The constructor is the function that has the same name as the class. Listing 7-2 contains the constructor for A3R3.

LISTING 7-2

```
1  public a3r3_p1() throws InterruptedException
2  {
3
4    //sets up the  sensors for a3r3
5      Vision = new UltrasonicSensor(SensorPort.S4);
6      Compass = new CompassHTSensor(SensorPort.S3);
7      SubstanceTemperature = new TemperatureSensor(SensorPort.S2);
8
9
```

```
10      //sets of the Tetrix Motor controllers with NXT
        //MicrontrollerPort
11       CF = new TetrixControllerFactory(SensorPort.S1);
12       MC = CF.newMotorController();
13       SC = CF.newServoController();
14       LeftMotor = MC.getRegulatedMotor(TetrixMotorController.MOTOR_1);
15       RightMotor = MC.getRegulatedMotor(TetrixMotorController.MOTOR_2);
16
17
18      //sets DC motors properly
19       LeftMotor.setReverse(true);
20       RightMotor.setReverse(false);
21
22      //sets up the robot arm and range of motion
23       Arm = SC.getServo(TetrixServoController.SERVO_2);
24       SC.setStepTime(7);
25       Arm.setRange(750,2250,180);
26       Arm.setpulseWidth(1500);
27
28      //sets up a pilot and navigator for the robot
29
30       WheelDiameter = 7.0;
31       TrackWidth = 34.5;
32       a3r3Pilot = newDifferentialPilot
                    (WheelDiameter,TrackWidth,LeftMotor,RightMotor);
33       a3r3Pilot.setTravelSpeed(6);
34       a3r3Pilot.setRotateSpeed(30);
35       Odometer = new OdometryPoseProvider(a3r3Pilot);
36       a3r3Navigator = new  Navigator(a3r3Pilot);
37       a3r3Navigator.singleStep(true);
38
39
40   }
```

Recall that as soon as the robot class is declared, the constructor is automatically called and is the first thing that your robot will execute. Notice in Listing 7-2 that for every major hardware component, there is an initialization process. Note also our naming conventions. We named things in software based on their hardware names. If you are using Tetrix-encoded motors or Tetrix servos, lines 10 through 26 in Listing 7-2 show you how to set them up for use in leJOS.

The Bluetooth Connection

The leJOS classes are divided into two sets when it comes to Bluetooth. There are the classes and code that run on the NXT microcontroller, and there are the classes and code that run on the computer. In our example, the computer is the initiator, and A3R3 and D1R1 are the receivers. There is a set of includes that are designed for the initiators/receivers for NXT microcontrollers and a different set of includes for the computer. Table 7-4 contains the imports by category.

TABLE 7-4   Important Classes that Must Be Imported When Using Bluetooth Communications

Bluetooth Imports for Computer	Bluetooth Imports for NXT Microcontroller	Classes for Computer and NXT Brick
`lejos.pc.comm.` `NXTCommLogListener`	`lejos.nxt.comm.` `Bluetooth`	`java.` `io.DataInputStream`
`lejos.pc.comm.NXTConnector`	`lejos.nxt.comm.` `BTConnection`	`java.` `io.DataOutputStream`
`lejos.pc.comm.NXTInfo`		`java.` `io.BufferedReader`
`lejos.pc.comm.` `NXTCommFactory`		`ava.` `io.InputStreamReader`
`lejos.pc.comm.` `NXTCommBluecove`		

The Stream classes in Table 7-4 replace the NXT-G and LabVIEW "Mailbox" concept. Listing 7-3 shows one of A3R3's communication methods. This method uses Bluetooth and is an example of how a receiver is set up to wait for a connection before executing any further code. The `waitForConnection()` method is a blocking method that will block until it gets a connection. It essentially stops the thread that is executing and waits. Unless your robot is multithreading, it will stop at the `waitForConnection()` call.

LISTING 7-3   One of A3R3's communication methods that uses Bluetooth.

```
1    public void communicate(float Heading,float Temperature) throws
     Exception
2    {
3
4        String connected = "Connected";
5        String waiting = "Waiting...";
6        String closing = "Closing...";
7        int CurrentColor =1;
```

```
 8      LCD.drawString(waiting,0,0);
 9      LCD.refresh();
10      BTConnection BlueConnection = Bluetooth.waitForConnection();
11      LCD.clear();
12      LCD.drawString(connected,0,0);
13      LCD.refresh();
14      DataInputStream Din = BlueConnection.openDataInputStream();
15      DataOutputStream Dout = BlueConnection.openDataOutputStream();
16      int n = Din.readInt();
17      LCD.drawInt(n,7,0,1);
18      Thread.sleep(2000);
19      Din.close();
20      Dout.writeFloat(Heading);
21      Thread.sleep(2000);
22      Dout.writeFloat(Temperature);
23      Thread.sleep(2000);
24      Dout.close();
25      LCD.refresh();
26      Thread.sleep(300);
27      LCD.clear();
28      LCD.drawString(closing,0,0);
29      LCD.refresh();
30      BlueConnection.close();
31      LCD.clear();
32
33
34   }
```

Notice that the `waitForConnection()` occurs on line 10. The input/output streams on lines 13 and 14 are used to pass data to and from the initiator. Once the connections are open, you can send and receive data as long as the input and output streams are open and the Bluetooth connection is open. This communication method reads in an `int` from the initiator on line 16, and notice on lines 20 and 22 that A3R3 sends its current heading and temperature measurement to the initiator. Listing 7-4 is the main function for A3R3.

LISTING 7-4

```
 1    public static void main(String [] args) throws Exception
 2    {
 3
 4
```

```
5        a3r3_p1 a3r3Robot = new a3r3_p1();//Constructor is called
6        a3r3Robot.moveArm(45);
7        a3r3Robot.moveArm(0);
8        a3r3Robot.rotate(-20);
9        Thread.sleep(1000);
10       a3r3Robot.navigatorStop();
11       LCD.drawString(a3r3Robot.distanceFromObject(),0,1);
12       a3r3Robot.moveArm(90);
13       Thread.sleep(6000);
14       a3r3Robot.moveArm(45);
15       a3r3Robot.communicate
                    (a3r3Robot.getDegrees(),a3r3Robot.substanceTemp());
16       a3r3Robot.rotate(40);
17       Thread.sleep(2000);
18       a3r3Robot.moveArm(0);
19       a3r3Robot.moveArm(90);
20       Thread.sleep(6000);
21
22       a3r3Robot.moveArm(0);
23       a3r3Robot.communicate
                    (a3r3Robot.getDegrees(),a3r3Robot.substanceTemp());
24       Thread.sleep(2000);
25       a3r3Robot.rotate(-20);
26       Thread.sleep(4000);
27       a3r3Robot.navigatorStop();
28       Thread.sleep(1022);
29       LCD.clear();
30
31
32
33
34
35   }
36
```

A3R3 uses the LEGO temperature sensor to take the temperature of two liquids. Figure 7-13 shows the LEGO temperature probe as it measures 2.5 in.

Once it has taken the first measurement, it uses the compass to get its current heading, and it sends the heading and the temperature to the initiator D1C3 (the computer). A3R3 rotates to the second liquid, takes the reading, and then sends the heading and the reading to the initiator. Once the initiator has the reading, the two readings will be compared, and the highest reading will be sent to D1R1. In this case, the goal is to have A3R3 send the information to D1R1, but A3R3 and D1R1 are not directly connected, so they use D1C3 as the middle man. Table 7-5 lists the basic steps to connect NXT bricks to a computer.

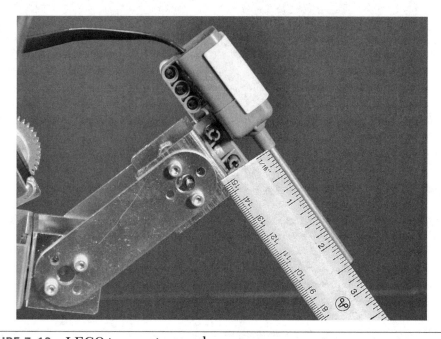

FIGURE 7-13 LEGO temperature probe.

TABLE 7-5 Basic Steps to Connect a Computer to an NXT Microcontroller (Computer Is the Initiator)

Steps	Bluetooth Connection from Computer to NXT Brick
1	Create NXT communication using `Comm Factory`.
2	Check whether creation of NXT Communication was successful.
3	Create `NXTInfo` object with address of Bluetooth device.
4	Check whether creation of `NXTInfo` object was successful.
5	Open a Communication connection using the Communication object.
6	Get the input and/or output stream from the Communication object.
7	Send and receive data.
8	Close the streams.

Listing 7-5 contains examples of these eight important steps. The listing also shows that the initiator will connect with three NXT microcontrollers:

- A3R3
- D1R1
- D1R2

So this little team consists of four connected devices.

LISTING 7-5   Initiator connecting to NXT microcontrollers and sending and receiving information

```
1    import java.io.DataInputStream;
2    import java.io.DataOutputStream;
3    import java.io.BufferedReader;
4    import java.io.InputStreamReader;
5    import lejos.pc.comm.NXTCommLogListener;
6    import lejos.pc.comm.NXTConnector;
7    import lejos.pc.comm.NXTInfo;
8    import lejos.pc.comm.NXTComm;
9    import lejos.pc.comm.NXTCommFactory;
10   import lejos.pc.comm.NXTCommBluecove;
11
12   public class initiator_master{
13      public static void main(String[] Args) throws Exception
14      {
15
16          String D1R1 = "D1R1";
17          String A3R3 = "A3R3";
18          String D1R2 =  "D1R2";
19          float Heading1 = 0;
20          float Heading2 = 0;
21          float Temperature1 = 0;
22          float Temperature2 = 0;   //Step 1
23          NXTComm  D1R1Comm = NXTCommFactory.createNXTComm
                                   (NXTCommFactory.BLUETOOTH);
24          NXTComm  A3R3Comm = NXTCommFactory.createNXTComm
                                   (NXTCommFactory.BLUETOOTH);
25          NXTComm  D1R2Comm = NXTCommFactory.createNXTComm
                                   (NXTCommFactory.BLUETOOTH);
26
27          if(D1R1Comm == null){   //Step 2
28
29              System.out.println("Could Not find Brick");
30              System.exit(1);
31          }
32          else{
```

```
33
34                    System.out.println("Found Brick" + D1R1);
35
36        }                       //Step 3
37    NXTInfo D1R1Robot = new NXTInfo
                          (NXTCommFactory.BLUETOOTH,D1R1,
                           "00:16:53:0C:54:91");
38    NXTInfo A3R3Robot =  new  NXTInfo
                          (NXTCommFactory.BLUETOOTH,A3R3,
                           "00:16:53:01:A9:8D");
39    NXTInfo D1R2Robot =  new  NXTInfo
                          (NXTCommFactory.BLUETOOTH,D1R2,
                           "00:16:53:14:1A:F3");
40    if(D1R1Robot == null   || A3R3Robot == null){   //Step 4
41
42        System.out.println("Could not create NXTInfo");
43        System.exit(1);
44    }
45    else{
46
47          System.out.println("Created NXTInfo: " + D1R1);
48
49    }
50    if(D1R1Comm.open(D1R1Robot,NXTComm.PACKET)){ // Step 5
51        System.out.println("Connected to " + D1R1);
52        System.out.println("Bytes to read" +
                              D1R1Comm.available());
53    }
54    else{
55          System.out.println("Did not Connect to " + D1R1);
56
57    }
58
59
60        if(A3R3Comm.open(A3R3Robot,NXTComm.PACKET)){
61
62        System.out.println("Connected to " + A3R3);
63        System.out.println("Bytes to read" +
                              A3R3Comm.available());
64    }
65    else
66
67          System.out.println("Did not Connect to " + A3R3);
68
```

```
69          }
70          if(D1R2Comm.open(D1R2Robot,NXTComm.PACKET)){
71              Voice.exec("festival --tts S-message4.txt");
72              System.out.println("Connected to " + D1R2);
73              System.out.println("Bytes to read" +
                                    D1R2Comm.available());
74          }
75          else{
76                  Voice.exec("festival --tts E-Message3.txt");
77                  System.out.println("Did not Connect to " + D1R2);
78
79          }
80
81          DataOutputStream Dout;
82          DataInputStream Din;
83
84          Dout = new DataOutputStream(A3R3Comm.getOutputStream());
            // Step 6
85          Dout.writeInt(8888);
86          Thread.sleep(OneSecond * 5);
87          Dout.close();
88          Din = new DataInputStream(A3R3Comm.getInputStream());
89          Heading1 = Din.readFloat();
90          System.out.println("Current Heading A3R3: " + Heading1);
91          Thread.sleep(OneSecond* 5);
92          Temperature1 = Din.readFloat(); //Step 7
93          System.out.println("Current Temperature from A3R3: " +
                                    Temperature1);
94          Thread.sleep(OneSecond *5);
95          Din.close();    //Step 8
96          if(A3R3Comm.open(A3R3Robot,NXTComm.PACKET)){
97          System.out.println("Bytes to read" + A3R3Comm.available());
98          }
99          else{
100
101             System.out.println("Did not Connect to " + A3R3);
102
103         }
104         Dout = new DataOutputStream(A3R3Comm.getOutputStream());
105         Dout.writeInt(8888);
106         Thread.sleep(OneSecond * 5);
107         Dout.close();
108
109         Din = new DataInputStream(A3R3Comm.getInputStream());
```

```
110        Heading2 = Din.readFloat();
111        System.out.println("Current Heading2 A3R3: " + Heading2);
112        Thread.sleep(OneSecond* 5);
113        Temperature2 = Din.readFloat();
114        System.out.println("Current Temperature2 from 3R3: " +
                               Temperature2);
115        Thread.sleep(OneSecond * 50);
116        Din.close();
117        float  HighestTemperature  = 0;
118        float  NewHeading  = 0;
119        if(Temperature1 < Temperature2){
120           HighestTemperature = Temperature2;
121           NewHeading = Heading2;
122        }
123        else{
124             HighestTemperature = Temperature1;
125             NewHeading  = Heading1;
126        }
127        if(D1R1Comm.open(D1R1Robot,NXTComm.PACKET)){
128           Voice.exec("festival --tts S-message4.txt");
129           System.out.println("Connected to " + D1R1);
130           System.out.println("Bytes to read" +
                              D1R1Comm.available());
131        }
132        else{
133             Voice.exec("festival --tts E-Message3.txt");
134             System.out.println("Did not Connect to " + D1R1);
135
136        }
137
138
139        Dout = new DataOutputStream(D1R1Comm.getOutputStream());
140        Dout.writeFloat(NewHeading);
141        Thread.sleep(OneSecond * 5);
142        Dout.writeFloat(HighestTemperature);
143        Thread.sleep(OneSecond * 5);
144        Dout.close();
145
146
147    }
148
149
150
151
152 }
```

Lines 23, 27, 37, 40, 50, 84, 93, and 95 show examples of the major steps in Bluetooth communication from and to the initiator or team leader when the initiator/ team leader is a computer and the team members consist of NXT microcontroller– based robots. Notice on lines 37 through 39 that we create connections for three NXT-based robots. Also notice that we talk to the robots one at a time. Each **read( )/ write( )** function of the streams is accompanied by a **Thread.sleep( )**. This gives the function time to finish. Unlike the **waitForConnection( )**, they don't necessarily block until they finish.

 It is possible to close( ) the connection while the robots are still in the middle of reading or writing a stream.

The **Thread.sleep( )** makes the threads a little more friendly to the flow of control. This simple program is a primer for our CSI program. We will go deeper into Java-powered robots in our CSI project.

Chapter 8

Got Linux and Darwin on Your Team of Robots?

The Lost Scrolls of Robotics: #8
Do you not see the logic of my plan?

—VIKI, *I Robot*

It's easy to get so totally immersed in the building and designing of your robots that such things as the robot's operating system and the operating system of the development environment have a tendency of falling off the radar. We know because we fall victim to this all the time. A couple of our members are frequently stuck in the design-build-admire-design cycle and often never end up programming the robots they build to do anything other than a few initial "Oh Wow!" tasks. But when we get right down to it, the operating system (OS) is both a gatekeeper to the power of the microcontroller and computer hardware and a silent partner on the team. The OS is absolutely critical to the success of our robot projects.

The Operating System as the Gatekeeper

Figures 8-1 and 8-2 show an NXT microcontroller motherboard and a Bluetooth communications module. These are the two main pieces of hardware that we are concerned with in this book. Without the NXT microcontroller, we have no robots. Without the Bluetooth module, we have no way to communicate among our team of robots. So it's natural to emphasize these pieces of hardware in our team robot designs. But remember that we do not manipulate this hardware directly. There are layers of software between us and the microcontroller or the computer's hardware. One of those layers of software is the OS. The OS provides an interface between programming languages and the hardware and between direct user manipulation and the hardware. If the OS does not provide an interface to a specific feature of the microcontroller, then we have no way to access that feature, plain and simple—

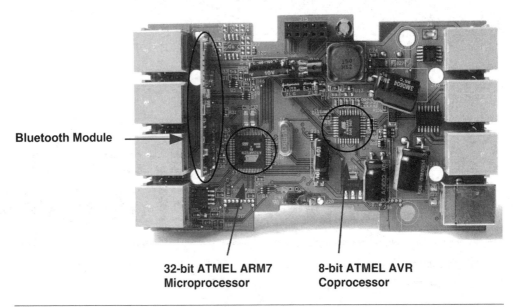

Bluetooth Module

32-bit ATMEL ARM7 Microprocessor

8-bit ATMEL AVR Coprocessor

FIGURE 8-1   NXT microcontroller motherboard.

unless, of course, we replace the OS with another OS that does provide access. We write programs for the microcontroller using languages such as LabVIEW and Java. But these languages are not the native language of the NXT controller or the AMD microprocessor. These languages require a chain of tools that convert the LabVIEW or Java code to ARM7, the language of the NXT microcontroller, or INTEL assembly, the

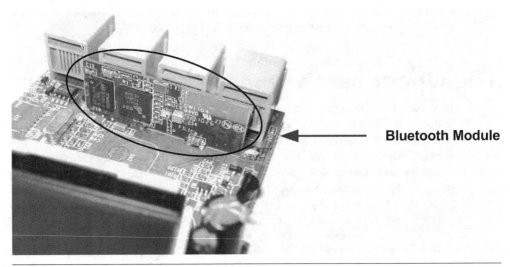

Bluetooth Module

FIGURE 8-2   Bluetooth communications module.

language of AMD and Intel microprocessors. And it's only possible to take advantage of the converted code if the OS exports an interface to the feature we are trying to program. The OS truly is the gatekeeper of the microcontroller and microprocessor.

Figure 8-3 shows the basic layering involved from your "bright-idea stage" to execution of commands by the robot. Notice that your bright ideas are pretty large. You then have to translate them into a programming language that hopefully captures the entirety of your genius! Look at the relationship in size between the programming language and the OS. Notice that the OS space is larger than the programming language space. This is so because in many situations the programming language might not give you access to all the capability that the OS is providing. Further, look at the ratio of the microcontroller space to the OS space. The OS space/firmware is the last piece of software that you really deal with. It may or may not provide you complete access to all the capability of the microcontroller. As you can see in Figure 8-3, computer language cannot give you any access to the microcontroller that is

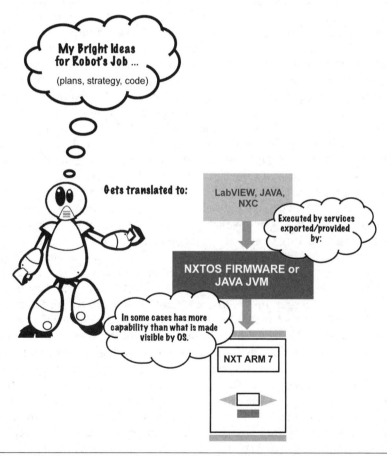

FIGURE 8-3   The basic layering between the "bright-idea stage" and the execution of commands by the robot.

not given to the computer language by the OS. Thus, if the OS was designed not to provide services for all the features that the microcontroller has to offer, typically you're stuck. This is why choice of OS is important both for the microcontroller and for the computer that you use for development. It acts as a gateway, either providing full access or limiting your options.

Operating System as Silent Partner

Not only does the OS act as a gatekeeper by allowing access to the hardware of the microcontroller and the microprocessors in your robot projects, but it also is a silent partner that handles many hardware tasks for you, such as managing device drivers, scheduling threads on the processor, loading programs into random access memory (RAM), and processing hardware interrupts. Programming your robot's microcontroller would require a lot more effort if it were not for the OS. In the case of Bluetooth communications, the OS, through its device drivers for the Bluetooth hardware, does the lion's share of the work. When we chose the OSs that we would use in building our team of robots, we had three primary goals in mind. First, we wanted an OS that provided the most access to the hardware and still was compatible with LEGO MINDSTORMS. Second, we wanted an OS that would offer the most flexibility and power as a silent partner on our robot teams. Third, we wanted an OS that supported all the development tools and development paradigms that we were used to. In this book we use two operating systems on the ARM7 of the NXT microcontroller (NXT-OS/firmware and leJOS JVM/firmware), and for the computer OSs we used primarily Linux Kernel 2.6, Dawrin Kernel 10.0 (Mac OSX), and Darwin Kernel 11.3 (Mac OSX). We also used Windows 7 for the Rubik's Cube solver, but our go-to OSs are Linux and Darwin. When we need maximum control, we go with Linux because we have full access to the source code of the OS and its device drivers. If there is some service that the OS does not provide, we can add it. If the device driver for some hardware is not available or does not behave the way we need it to, Linux allows us to modify or add the functionality we need.

Ultimately, the OS that you use on your projects is up to you and will be based on how much power and flexibility are required by your team of robots. Although we do provide code for projects in NXT-G, LabVIEW, NXC, and Java in this book, Java is our primary programming environment for the teams of robots that we program in our laboratory. We have found that the leJOS/firmware offers us (in the long run) the most access to the ARM7 microcontroller of the NXT brick, and the Java environment supports full agent-oriented programming as well as object-oriented programming. These are two of the most powerful paradigms for programming robots available. In addition to supporting these programming methodologies, Java is the language of many mobile devices and embedded computers and one of the de facto languages of the Internet. Using the Java environment over Linux or Darwin provides us access to a world of software libraries, Internet communication capabilities, mobile devices, development tools, and so on.

Computer-Aided Design (CAD) Software for Your Robot Designs Using Digital Designer

Computer-aided design (CAD) is one of the most important pieces of software used during the design of your robots. CAD software allows you to visually select the parts you are going to use and visually build them into the robot design that you have in mind. Thus, prior to touching any hardware or robot parts, the CAD software allows you to experiment with various designs. For the projects in this book, we used two CAD packages: LEGO Digital Designers and LCAD. LCAD was used for our Rubic's cube solver, and we used LEGO Digital Designer for the remainder of the projects. Figure 8-4 is a screen capture of the LEGO Digital Designer (Version 4.2).

Figure 8-4 also shows a MINDSTORMS robot design. Notice on the left-hand side of the figure that you can see some of the MINDSTORMS parts that can be used to

FIGURE 8-4   LEGO Digital Designer Version 4.2.

build your robots. This is one of the important features of a CAD package. It will have visual representations of all the parts that can be used in constructing your robots. While the LEGO Digital Designer is not supported in Linux, it is supported in Darwin/MacOSX. Another critical feature of the LEGO Digital Designer software is the "Build Instruction" feature. This feature displays type and number of parts needed for each step in the instruction. You can then use these build instructions to actually construct the robot that you designed visually or share the build instructions with others who would like to construct your design. Figure 8-5 show examples of the Digital Designer's "Build Instruction" guide for the simple robot that is shown in Figure 8-4. The LEGO Digital Designer can be obtained from http://ldd.lego.com/en-us/download/.

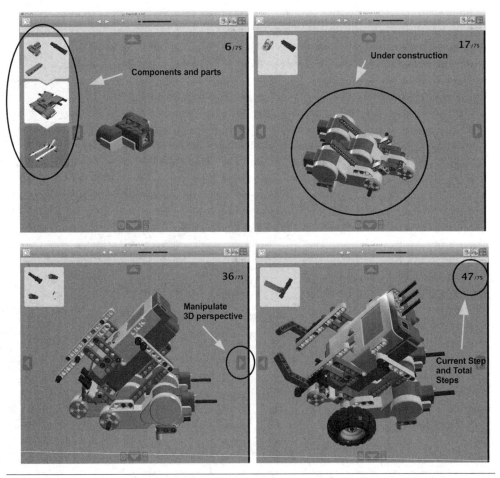

FIGURES 8-5 Some build instructions for the simple robot in Figure 8-4.

Development Languages for Programming Your Robots

NXT-G and LabVIEW both run under MacOSX, so if you wish to program your MINDSTORMS robots using graphical language G in a Darwin environment, you can. LabVIEW is also available for Linux, but at the time of this writing, there is no MINDSTORMS toolkit for Linux, and NXT-G does not run directly under Linux. So both graphics environments are a no-go for Linux. Both Linux and Darwin do have support for NXC programming, though.

The Simple NXC (Almost C) Tool Chains

We show samples of NXC programming in Chapter 1. NXC allows you to program your robots in a language that is very close to the C programming language. So, if you are familiar with C programming, you will feel at home using the NXC programming environment for either Linux or Darwin. There is a simple tool chain for NXC in the Linux/Darwin environment. It consists of a couple of command-line tools. The command-line tools for the NXC environment run under Linux/Darwin. To compile your program, use

```
nbc your_program.nxc -O=your_program.rxe
```

To download your program, use

```
nbc -d your_program.rxe (make sure you use the program with the .rxe
suffix)
```

nbc has several important options that can be accessed using the nbc - help option:

```
-S=<portname>: specify port name (usb), brick resource name, or alias
   -d: download program
   -r: download and run program
   -b: treat input file as a binary file (don't compile it)
   -q: quiet
   -n: prevent the system file from being included
   -D=<sym>[=<value>]: define macro <sym>
   -x: decompile program
   -Z[1|2]: turn on compiler optimizations
   -ER=n: set maximum errors before aborting (0 == no limit)
   -PD=n: set max preprocessor recursion depth (default == 10)
   -O=<outfile> : specify output file
   -E=<filename> : write compiler messages to <filename>
   -I=<path>: search <path> for include files
   -nbc=<filename> : save NXC intermediate NBC code to <filename>
   -L=<filename> : generate code listing to <filename>
   -Y=<filename> : generate symbol table to <filename>
```

```
-w[-|+] : warnings off or on (default is on)
-sm[-|+] : status messages off or on (default is on)
-EF : enhanced firmware
-safecall: NXC will wrap all function calls in Acquire/Release
-api: dump the API to stdout
-v=n: set the targeted firmware version (default == 128, NXT 1.1 == 105)
```

This is just about the only place you can find these options, so make note of the simple command **nbc -help**. The **nbc** development environment for MacOSX (Darwin) and Linux can be obtained from htttp://bricxcc.sourceforge.net/nbc.

NXC is a powerful development environment. It gives you a wide range of access to the NXT microcontroller. The biggest problem that you will have if you decide not to use NXT-G to program your robots is the compatibility of the language you choose with all the available sensors. If you are only using stock sensors that come with MINDSTORMS robot kits, you're usually safe with NXC, LabVIEW, or leJOS (Java). However, if you start getting into the Vernier probes or some of the less popular HiTechnic sensors, there may not be support in leJOS (Java). In such cases, you may find yourself writing a lot of low-level code to get your sensors to work. We only encountered this situation once for the projects that are shown in this book, and that was with the Vernier pH probe. Although leJOS did not offer easy out-of-the-box access, it was possible with a little extra programming to get the Vernier pH probe working with the MINDSTORMS microcontroller.

Using Eclipse in the Linux/Darwin Environments

If your preference in a Linux/Darwin environment is C programming, the NXC development environment is the best choice. If you prefer Java with support for object-oriented programming and agent-oriented programming, as well as easy access to other mobile and embedded devices, then leJOS (Java) is hands-down the best way to go. We covered the power of Java and your robots in Chapter 7. Eclipse is one of the easiest tools to use with leJOS. Eclipse is a graphical development environment that supports Java and C++ and has a built-in plug-in for leJOS. Figure 8-6 is a screenshot of the major view of the Eclipse environment.

Once you have Eclipse installed, the first thing you need to do is get the LEGO MINDSTORMS plug-in for Eclipse. This plug-in can be downloaded from http://lejos.sourceforge.net./tools/eclipse/plugin/nxj. To install this plug-in select the "Help: Install New Software" option in the Eclipse environment (see Figure 8-7). Follow the dialog instructions of that screen. Once the plug-in is installed, you will be able to update the NXT firmware from the built-in NXT firmware to leJOS (but not the other way around). You cannot go back to the original firmware from inside Eclipse. If you need to go back to the original firmware, use the NXT-G "Environment/Tools" option. Updating firmware is a menu option when the Eclipse plug-in is installed (see Figure 8-8).

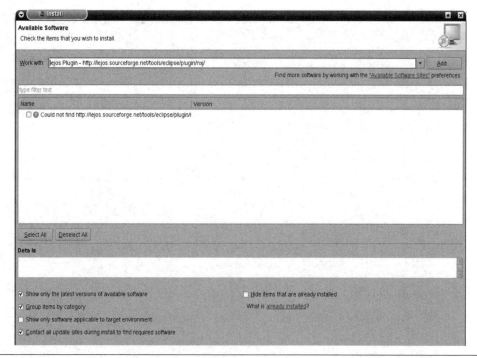

FIGURE 8-6   The main view for Eclipse.

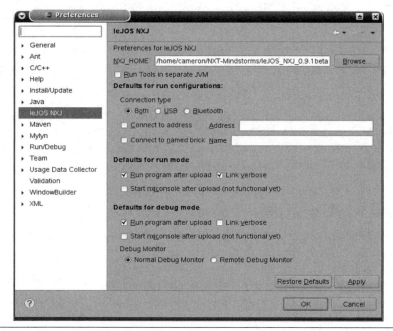

FIGURE 8-7   The "Install New Software" option screen for Eclipse.

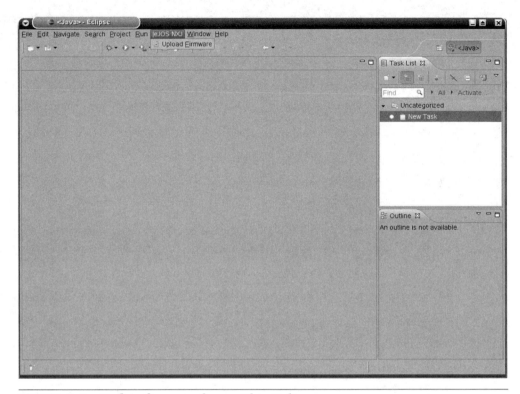

FIGURE 8-8 Update firmware from within Eclipse.

In addition to direct support for Flashing leJOS firmware, the Eclipse environment also supports the creation of leJOS projects that are meant to run on either the NXT brick or the PC. Figure 8-9 shows where the project-creation support is in the Eclipse environment.

Select "New," and then pick "Project" (not "Java Project"). You then will see the two options for leJOS projects, as shown in Figure 8-9. It's easy to overlook the role the OS plays in developing code for your robots, but just remember it plays a significant role.

What About My Files? (Where Do They Go?)

Keep in mind that the NXT brick is limited in storage. Depending on the size and complexity of the programs that you write for your robots, you will have room for only a few programs. Where are extra programs stored? What if you've developed dozens of programs for your robot or team of robots, now what? Well, you can use your computer's OS to store the files for you, and depending on which OS you select, there are nice features that let you organize and retrieve files in almost any way you choose. You can organize your files by project, by robot, or by type using the graphical

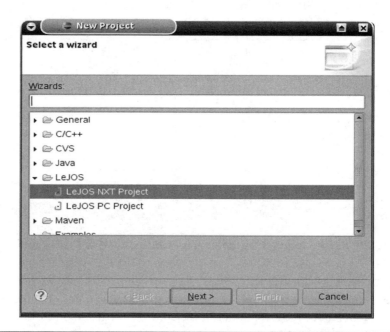

FIGURE 8-9   The project-creation support in the Eclipse environment.

user interface of your OS. For example, one of the organizations for the projects that we used in this book is shown in Figure 8-10.

In some cases you even may want to break down the files based on robot names. The important thing to remember is to be consistent and use a naming technique that will help you to remember what's what, because once you've developed and gotten a program working and uploaded to the robot, it may be some time before you have to make changes or add functionality to your robot's program. The way you organize and store the robot's files will help your memory kick in when the time comes.

Linux and Darwin as Runtime Environments

The OS you choose will depend on how much programming you do and the devices you want to program. The only device we are interested in writing programs for is the MINDSTORMS NXT controller. In this case, Darwin/Linux/Windows all offer about the same functionality when it comes to leJOS (Java) or NXC (almost C). MacOSX and Windows have a definite advantage over Linux when it comes to programming in the graphical language G using either LabVIEW or NXT-G. Also, MacOSX and Windows have the advantage over Linux when it comes to CAD tools such Digital Designer, LCAD, and LDraw. So your choice of CAD tools also may influence your choice of OS. In some cases, though, we have to program not only our robots but also other mobile or embedded devices or other computer applications. These situations cause OS

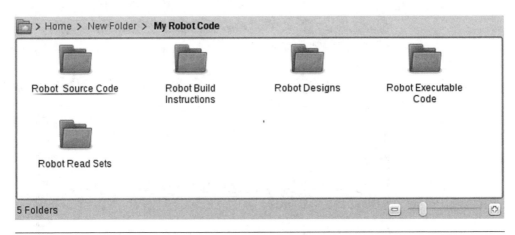

FIGURE 8-10 Using the graphical user interface (GUI) to organize your files.

selection to be not so simple. In addition to a development environment, the OS also has tremendous responsibilities in the runtime environment. And depending on the sophistication of the applications that your team of robots have to execute, the OS can be your friend or foe.

Runtime Capability When the Computer Is the Team Leader

In several of the projects in this book we use the computer as the initiator or team leader of the robot team. When the computer is the team leader, the entire team has access to any piece of hardware, data, computer language, communications protocol, and so on that the computer has access to. For example, our NXT controller has a single 32-bit ARM7 microprocessor. On the other hand, the computer that is connected to the BRON with our other NXT bricks has four microprocessors that each run approximately 3 GHz, with 16 GB of RAM and a terabyte of storage! In addition to this, the computer that is the initiator in our MINDSTORMS BRON is itself connected to a supercomputer cluster. This means that indirectly any of our NXT bricks during runtime has access to a full-blown supercomputer cluster. Wow! Of course, for the projects that we present in this book, we did not need to access the computers that our initiator computer had access to, but it's nice to know that the extra power is there when you need it. Simple access to the cluster is provided because of the basic design of the Linux OS. Linux and Darwin have multithreading and parallel-processing capability that manages the Bluetooth communications in a very robust way. Further, the more powerful the OS, the more tasks it can do simultaneously. Both Linux and Darwin have been designed from the very beginning with multitasking, multiprocessing, and even parallel processing in mind. In addition to just the sheer power that an OS can bring at runtime, we have to look at all the available software that may be running on that computer at runtime that your team of robots may take advantage of. As we have seen when it comes to available software, not all OSs are created equal. But it's available software at runtime that we are most concerned with. If the computer is a member of the team, then whatever the computer has access to, your entire team (through Bluetooth communications) has access to. For example, in our crime scene investigation (CSI) project, we need to analyze an unknown substance that our security robot located in our warehouse. We have an artificial intelligence program (written in the obscure language Prolog) that has the ability to do such an analysis. The problem is that the NXT brick does not have enough horsepower to run the program in a timely fashion, and even if the process were fast enough, the NXT brick doesn't have enough memory. To add insult to injury, the NXT brick does not understand the obscure language Prolog. So how was the magic performed?

- **Step 1:** The NXT brick firmware was updated to run leJOS.
- **Step 2:** Linux/Darwin-based computers used Java to talk to the NXT bricks.
- **Step 3:** JIP Prolog runs as a Java program on Linux/Darwin.

Figure 8-11 gives you an overview of the logical connection and shows how NXT bricks can have access to anything that the computer's OS system has access to.

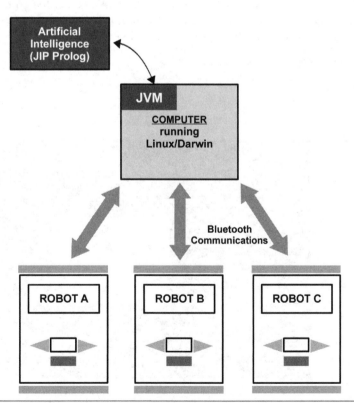

FIGURE 8-11   NXT bricks' logical access to everything through the computer's OS.

The BlueZ Protocol Can Handle NXT Bricks

Bluetooth connections under Linux are handled with the BlueZ protocol stack. The BlueZ protocol is also supported on Darwin. If you're running a recent Linux distribution, it is probably already installed on your system. If not, you can download it from http://bluez.sourceforge.net/. For most of the Bluetooth connections using Java, we were using SuSE Linux 11.0 and 11.1. We used the micro-Bluetooth adapter from Rocket Fish. This adapter worked right out of the box with both SuSE 11.0 and 11.1. Figure 8-12 shows a photo of the Bluetooth adapter from Rocket Fish.

The good news about BlueZ is that it recognizes and communicates with the NXT Bluetooth hardware and software. In addition to communicating easily with the NXT microcontroller, we used it in a couple of projects in this book to communicate with the Android smart phone. Thus, once you're familiar with the basic Bluetooth tools under Linux, you can quickly set up a team that consists of NXT bricks and Android smart phones with a Linux computer as the initiator. Although there are GUI tools that can be used to work with Bluetooth under Linux, we suggest that you first familiarize yourself with the three basic command-line tools. While not all Linux distributions will have the same GUI tools, they typically will support the command-line tools `hciconfig`, `hcitool`, and `rfcomm`. In addition to these three command-line utilities, there are two configuration files that are involved:

```
/etc/bluetooth/rfcomm.conf
/etc/bluetooth/hcid.config
```

FIGURE 8-12 A photo of the Bluetooth adapter from Rocket Fish.

Both files can be configured using the appropriate utility: `rfcomm` or `hciconfig`. To get a detailed description of what the `rfcomm` and `hciconfig` utilities do, type the following:

`man rfcomm` or `man hciconfig`

Listing 8-1 shows an example of our `rfcomm` file and how it is set up to work with a couple of NXT bricks.

LISTING 8-1

```
rfcomm0 {
#       # Automatically bind the device at startup
        bind yes;
#
#       # Bluetooth address of the device
        device 00:16:53:0C:54:91;
#
#       # RFCOMM channel for the connection
        channel    1;
#
#       # Description of the connection
        comment "D1R1";
}
rfcomm1 {
#       # Automatically bind the device at startup
        bind yes;
#
#       # Bluetooth address of the device
        device 00:16:53:01:A9:8D;
#
#       # RFCOMM channel for the connection
        channel    1;
#
#       # Description of the connection
        comment "A3R3";
}
```

This file contains one `rfcomm` description for every device with which you will be communicating. They are named `rfcomm0`, `rfcomm1`, `rfcomm2`, and so on. Note that the comment line has the names of the two bricks. Note also that `rfcomm0` is set up for D1R1, and `rfcomm1` is set up for A3R3. We have it set up so that the computer and these robots are automatically paired whenever we start the computer up. If this is not the behavior you want, then change the value of Bind from **YES** to **NO**. Also notice this in both device lines in Listing 8-1.

The Bluetooth address of the device is

```
device 00:16:53:0C:54:91;
```

and the Bluetooth address of the other device is

```
device 00:16:53:01:A9:8D;
```

These are the addresses of the two NXT robot microcontrollers (bricks), D1R1 and A3R3. When working with computer-based Bluetooth or Android, you will need these addresses. In this case, we have the channel set to 1 for both NXT bricks. This is a logical channel, and we can assign a different channel to each Bluetooth application. See the main pages for more details on the channels. Just to keep things simple for the examples in this book, we assign all the devices on the team to channel 1. The `rfcomm.conf` file is located in the `/etc/bluetooth` directory. The second configuration file that is used with Bluetooth is `hcid.conf`. Listing 8-2 shows the `hcid.conf` file that we used for several of the projects in this book.

LISTING 8-2

```
#
# HCI daemon configuration file.
#

# HCId options
options {
    # Automatically initialize new devices
    autoinit yes;

    # Security Manager mode
    #    none - Security manager disabled
    #    auto - Use local PIN for incoming connections
    #    user - Always ask user for a PIN
    #
    security user;

    # Pairing mode
    #    none  - Pairing disabled
    #    multi - Allow pairing with already paired devices
    #    once  - Pair once and deny successive attempts
    pairing multi;

    # Default PIN code for incoming connections
```

```
    #    passkey "BlueZ";

}

# Default settings for HCI devices
device {
    # Local device name
    #    %d - device id
    #    %h - host name
    # name "BlueZ %h (%d)";
    name "krell %h (%d)";

    # Local device class
    class 0x000100;

    # Default packet type
    # pkt_type DH1,DM1,HV1;

    # Inquiry and Page scan
    # valid parameters: enable | disable
      iscan enable;
      pscan enable;

    # Default link mode
    #    none   - no specific policy
    #    accept - always accept incoming connections
    #    master - become master on incoming connections,
    #    deny role switch on outgoing connections
    lm master;

    # Default link policy
    #    none    - no specific policy
    #    rswitch - allow role switch
    #    hold    - allow hold mode
    #    sniff   - allow sniff mode
    #    park    - allow park mode
    lp rswitch,hold,sniff,park;
}
```

Notice in this listing that you can set options such as the PIN of the devices to which you're connecting. You can decide whether the computer will be a master or not for the team. You can set pairing mode and security. This file is also where you can change the name of the computer as seen by the NXT bricks. Although these files can be edited and changed manually, it is usually easier to use the command-line tools or the GUI tools if your Linux Distribution supports them.

To make sure our Bluetooth adapter is up and operational, we type:

```
hciconfig hci0 reset up
```

To verify that our computer is seeing our NXT bricks, other computers, or our Android smart phone, we can use a **hcitool** scan. Figure 8-13 shows what we received when we execute a **hcitool** scan.

Notice that we pick up an NXT and our Android. First, the scan shows you what devices are "On," discoverable, and within range of the computer. Second, it shows you the Bluetooth addresses of those devices. These addresses are helpful in setting up automatic pairing and so on in **RFCOMM**. We also used KBluetooth in our projects. KBluetooth is a GUI tool that allows you to pair, manage, and configure your Bluetooth devices. If the command-line approach feels a little dicey and you're lucky enough to have KBluetooh involved, it makes things a lot easier. Figure 8-14 show KBluetooth configuration, monitoring, pairing, and screens.

KBluetooth runs under the KDE environment. If you are using Gnome or one of the other desktops, they will have another GUI utility for configuring Bluetooth. Once your Bluetooth configuration files are set and you run **hciconfig hci0**, you should be able to access the NXT bricks, other computers, and the Android smart phone. The code **hci0** is the logical name of your Bluetooth adapter. And the **re-set up** command is a parameter to the **hciconfig** utility that is directing it to turn Bluetooth on.

In this book we focus on building a BRON using a team of NXT-based robots and other Bluetooth-capable devices. We would like to emphasize here that when one of the members of the team is a computer, then the potential for what the team of robots can accomplish rises exponentially, and the possibilities are almost endless. The more powerful, flexible, and open the OS is that runs the computer, the more power

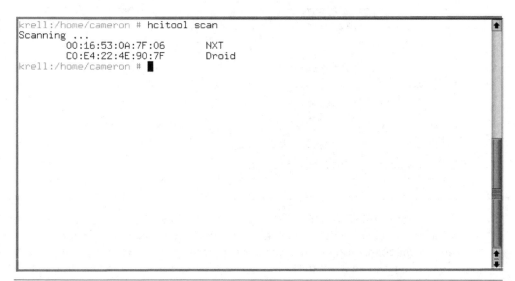

FIGURE 8-13 The output for a **hcitool** scan.

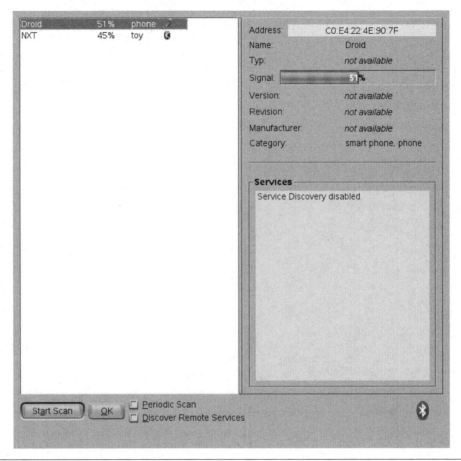

FIGURE 8-14   KBluetooth configuration, monitor, pairing, and screens.

your team of robots will have access to. We include Linux and Darwin (MacOSX) here because in practice we have found no barriers. To the contrary, the open-source tradition of Linux and the source-code traditions of UNIX have allowed us to modify, correct, and customize OS components that other OSs consider proprietary and off limits. Thus, when combining MINDSTORMS NXT robotic kits with Linux or MacOSX (Darwin), we were limited only by our imaginations and, of course, budgets.

Chapter 9

Advanced Teamwork: One for All!

The Lost Scrolls of Robotics: #9

We will add your biological and technological distinctiveness to our own.

—Locutus of Borg, *Star Trek: The Next Generation, Episode 26, Part 1: The Best of Both Worlds*

Recall that one of our primary reasons for putting our robots into teams is that no one robot can do it all. There is no general-purpose, one-size-fits-all robot design. Because microcontrollers have limited sensor capability, limited actuator capability, limited power, and so on, it is sometimes necessary to recruit fellow robots to get the job done. In these cases, enlisting a few team members will get the job done. But sometimes the answer is not to add more robots. Some tasks scream out for a single robot but with more horsepower than a single robot may have. In this case, we expand our team concept to include the notion of a *collective*, that is, a collection of two or more microcontrollers combined together to make a *single multicontroller/multiagent robot*. The collective, or hive, robot is housed in a single physical framework. It may be stationary or mobile. If it is mobile, then when one controller moves, they all move. Whereas in the typical team of robots each robot has a separate frame, each robot in a collective is physically distinct from the other robots on the team. The collective, in this case, has *one robot body*, so to speak. The need for a collective design becomes obvious when it is clear that a single robot build can handle the sensory needs or the physical manipulation tasks, but the NXT microcontroller doesn't have enough memory to do the job on its own or does not have enough processing speed, power, input-output (I/O) ports, and so on. So we combine two or more processors to get the job done, resulting in one physical robot body/framework with multiple microcontrollers.

If It Works for Me, It'll Work for You

All of the communications and coordination techniques that we have discussed so far still apply to collectives. Recall from Chapter 1 that we have

- Time chronology/synchronization
- Event-based communication/synchronization
- Message-based communication/synchronization

Each of these methods can be used alone or in any combination to coordinate and synchronize the activity within a collective.

In this project, we build a collective from two NXT microcontrollers and an Android-based smart phone. These are basically the components of a three-robot team, but instead of three separate robot frames, we construct one robot within a fixed frame whose sole purpose is to solve the famous Rubik's Cube. All communication and coordination will be directed at physically manipulating the cube itself as well as analyzing which colors are where and how to get them where they need to be.

This is a scenario where too many cooks spoil the stew. We only need two end effectors to manipulate the cube. We really only need one set of eyes to see what colors are where. Building multiple robots in this case would be overkill. But the processing power and memory requirements of solving the Rubik's Cube using our methods are beyond the capabilities of a single NXT microcontroller. This is what makes the *collective* or *hive* design a natural choice.

Designing a Rubik's Cube–solving LEGO robot has long been one of our goals, dating back to the time of the original RCX-based MINDSTORMS kits. It brings together several of our group's personal interests—designing LEGO contraptions, puzzle solving, and mathematics—into a project that has a very high "coolness factor."

From Team to Collective and Back

We start out with individual robots, and by adding a team mode and possibly a little repurposing, we allow multiple robots to work as one complex robot to accomplish a goal or solve a problem. The collective or hive design allows one robot to work as many while accomplishing a single goal, performing a single task, or solving a single problem. Both approaches accomplish the same thing—teamwork!

The Collective

The collective consists of two NXT components—Twisty and Flippy—and one Android-based smart phone. There are two main reasons for employing a team of robots rather than just one:

- A total of five servos are used in the project; an NXT microcontroller can control only three servos without a special multiplexer.
- Each device has a specialized set of capabilities that complement each other in the overall design.

The three devices have two methods by which they all can communicate: either connect to and talk through a computer or use Bluetooth communication. Bluetooth is adequate because no computer is required (other than to develop the programming).

Dividing Up the Labor

The two NXT units each have 32-bit ARM processors running at 48 MHz, with 64 kB of RAM and 256 kB of Flash storage. The capabilities of the Android device will vary, but the processing speed and memory capacity are certainly orders of magnitude beyond the NXT units. Of main importance is the amount of available storage because the lookup tables consume a large amount of space. The Android device is the only one of the three computers that has sufficient storage; thus the solver's tables must be stored there. The solver itself could be located on any of the devices because its code is rather small. The advantages and drawbacks of each location are based on communications overhead. If the solver is on one of the NXT units, it has the advantage of controlling that NXT unit's functions (i.e., gripping, flipping, or twisting) without the need for Bluetooth communication. However, it must exchange data with the Android device for each twist because it must request the twist information from the Android's lookup tables. It also must communicate with the other NXT unit; this requires a nontrivial amount of time because switching communications between the two other devices can take 0.5 s on an NXT unit.

The better approach is to locate the solver on the Android device. Although it requires communicating commands to both the NXT units, the Android is better suited to switch devices during communication. The total amount of communication (commands to the NXT units and table lookup information) is roughly the same for either option. The devices and the tasks that each perform are listed in Table 9-1. Figure 9-1 shows the relationship between the three components (Android and the two NXT units) for the Rubik's Cube solver robot.

TABLE 9-1   List of Devices and Their Basic Tasks Used by the Rubik's Cube Solver Robot

Device	Tasks
Android device	Photographing the cube
	Identifying facelet colors
	Code and tables for the solver
	Sending specific commands to Twisty and Flippy
NXT Twisty	Turning the twisters jointly or individually
	Aligning the cube faces
NXT Flippy	Gripping and releasing the cube
	Using the slider and gripper in combination to flip the cube

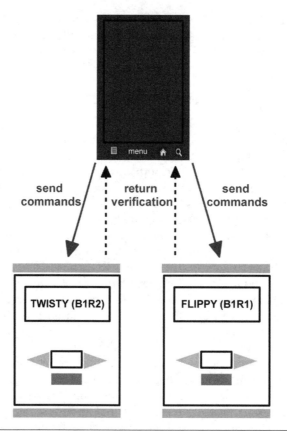

FIGURE 9-1 Relationship between the Android and the two NXT units of the Rubik's Cube solver robot.

The Flippy NXT controls both servos involved in flipping the cube, whereas the Twisty NXT controls the servos that turn and align the faces. This minimizes communication between the devices because a single command—FLIP, TWIST, etc.— can be sent to a single NXT, which then coordinates the low-level instructions to the servos. Figure 9-2 shows the main components for the Rubik's Cube solver robot.

Communicating with Flippy and Twisty

The overall communication scheme is a basic master-slave configuration. The Android initiates communication with the NXT units, and all communication originates from the Android. Each NXT unit listens to the Android for commands and sends back a simple reply when the command is completed. The Android sends 2-byte commands to the NXT units. The first byte is the actual command, whereas the second byte is a pseudorandom number used as a check byte for communication integrity. The NXT units receive the command and execute it using an if-else-if

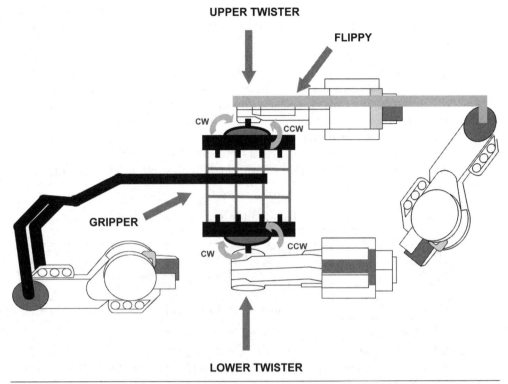

FIGURE 9-2 The main components for the Rubik's Cube solver robot.

code chain. They return a 2-byte packet back to the Android. The first byte is a status code, whereas the second byte is the received check byte plus one. To complete a command, the Android checks both bytes for a successful command completion and valid check byte. The six commands that the Android can send to Flippy are listed in Table 9-2. Receipt of any other command is ignored, returning **ERROR** to the Android.

TABLE 9-2 The Six Flippy Commands Sent by the Android

Command	Purpose	Returns
PING	Acknowledge that communication link is active	OKAY
DONE	Shut off motors but maintain communication	OKAY
SHUTDOWN	Shut off motors and quit	OKAY
GRIP	Extend the gripper	OKAY
RELEASE	Retract the gripper	OKAY
FLIP	Flip the cube	OKAY

The 16 commands that the Android can send to Twisty are listed in Table 9-3. Receipt of any other command will be ignored by Twisty and an ERROR code will be sent back to the Android. The PING command is sent to both NXT units by the Communicator module after successful Bluetooth connection to verify that the links are open and active. If either of the PING commands fails, the SHUTDOWN command is sent to both NXT units, and the Android app finishes. The DONE command is sent by the UI thread by pressing the "Stop" button. The SHUTDOWN command is sent either from the UI thread by pressing the "Stop" button or by various modules when an error is detected. The remaining commands are sent by the Manipulator module. The Manipulator takes high-level twist information (e.g., "Twist orange face clockwise") and turns it into a sequence of low-level commands for Twisty and Flippy.

TABLE 9-3   The 16 Twisty Commands Sent by the Android

Command	Purpose	Returns
PING	Acknowledge that communication link is active	OKAY
DONE	Shut off motors but maintain communication	OKAY
SHUTDOWN	Shut off motors and quit	OKAY
TWIST_TOP_CW, TWIST_TOP_CCW, TWIST_TOP_HALF	Twist top face clockwise 90 degrees, counterclockwise 90 degrees or 180 degrees	OKAY for each
TWIST_BOTTOM_CW, TWIST_BOTTOM_CCW, TWIST_BOTTOM_HALF	Twist bottom face clockwise 90 degrees, counterclockwise 90 degrees or 180 degrees	OKAY for each
SLICE_TOP	Twist top face clockwise 90 degrees and twist bottom face counterclockwise 90 degrees	OKAY
SLICE_BOTTOM	Twist top face counterclockwise 90 degrees and twist bottom face clockwise 90 degrees	OKAY
SLICE_HALF	Twist top and bottom faces 180 degrees	OKAY
ANTISLICE_CW	Twist top and bottom faces clockwise 90 degrees	OKAY
ANTISLICE_CCW	Twist top and bottom faces counterclockwise 90 degrees	OKAY
ALIGN_ON	Perform alignment after subsequent twists	OKAY
ALIGN_OFF	Do not perform alignment after subsequent twists	OKAY

Solving a Rubik's Cube

Let's talk about the algorithm used by the Rubik's Cube solver to unscramble a cube. If math makes you squeamish (hopefully not!), then you might want to skip over the "Phase Details" section. We'll start with a brief overview of a Rubik's Cube and the terms used in the rest of the chapter and then describe the process and the mathematics behind the process.

Remember the Cube: Parts of the Cube

For this project the *robot environmental attribute description* (READ) set will describe the cube itself. The only object this robot will encounter or interact with is the Rubik's Cube. Table 9-4 is the READ set for the Rubik's Cube solver. Figure 9-3 shows the parts, pieces, and faces of the Rubik's Cube. A standard 3- × 3-in Rubik's Cube is made up of 21 pieces:

- A center section consisting of a hidden piece that connects the center sections of each of the six cube faces

TABLE 9-4   The READ Set for the Rubik's Cube Solver

Attribute	Value
Object: Physical Workspace	
Width	5.7 cm
Length	5.7 cm
Height	5.7 cm
Shape	Cube
Surface	Flat, smooth
Faces	6
Number of parts	54 cubies
Object: Color (Light)	
Number of colors	6
Colors	Red, orange, blue, green, white, yellow
Object: Pieces	
Corner	8 with 3 exposed faces
Edge	12 with 2 exposed faces
Center	Hidden center piece connects center pieces
Moving parts	3 layers (on X, Y, Z axes) that turn 360 degrees

- Eight corner pieces with three differently colored stickers, one on each of its exposed faces
- Twelve corner pieces with two differently colored stickers, one on each of its exposed faces

The 20 smaller cubes comprising the corners and edges are commonly called *cubies*; the 54 colored sections comprising the six cube faces are known as *facelets*. Each of the six faces can be rotated 90 degrees; this shuffles 20 of the 54 facelets. A cube is scrambled by performing some random sequence of twists on the cube faces. The cube is then *solved* by performing twists to restore each face to a solid color (each side having nine facelets of the same color).

The faces of a standard Rubik's Cube are colored in a specific manner:

- White and yellow are on opposite faces.
- Red and orange are on opposite faces.
- Green and blue are on opposite faces.
- If the cube is sitting on a flat surface with the white face on top, then the orange face is to the right of the blue face.

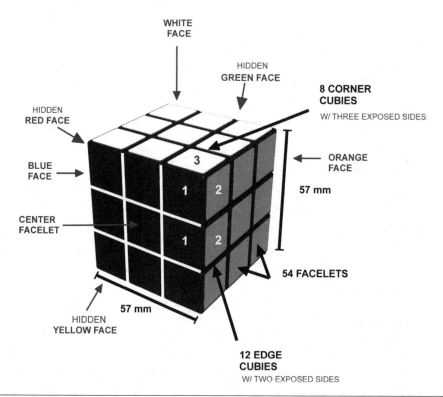

FIGURE 9-3 The parts, faces, and facelets of a standard 3- × 3-in Rubik's Cube.

Each face can be identified by the color of its center facelet because the center facelets do not move relative to each other.

Normally, twists of the faces are indicated in a color-independent manner. The universally accepted notation is *Singmaster notation*, named after David Singmaster.[*] This notation uses six letters—U, D, F, B, L, and R—to denote a clockwise quarter turn of the top (*up*) face, bottom (*down*) face, *front* face, *back* face, *left* face, and *right* face, respectively, and

- An apostrophe following one of the previous letters, denoting a counterclockwise turn of the given face
- The number 2 (sometimes written as a superscript, as in B^2) following one of the face letters, denoting a half turn of the given face.

Unfortunately, Singmaster notation will not work well in this project. The robot can only turn the top and bottom faces, and the cube is flipped during the solution process to place a face to be twisted on either the top or bottom of the cube. To handle this, we will use a modified Singmaster notation; instead of directions, we will use W, B, O, G, R, and Y to denote the color of the face to be twisted. We will also add two letters to the notation:

- An S after one of the colors denotes a *slice* move, which turns the given face clockwise one-quarter turn, and the opposite face is turned counterclockwise a quarter turn. The net effect is to turn the middle slice a quarter turn counterclockwise (hence the move's name).
- A slice move can be *squared*, as above, to denote a half turn of the center slice.
- An A after one of the colors denotes an *antislice* move, which performs a clockwise quarter turn on both the given face and the opposite face.
- An antislice can be *clockwise* or *counterclockwise*, the latter using an apostrophe following the A.

Examples of these additional twists are given in the next two figures. Figure 9-4 shows the effect of the two slice moves on a solved cube. Figure 9-4a shows a slice, and Figure 9-4b shows a squared (or half) slice. Figure 9-5 shows the two antislice moves. Figure 9-5a shows a clockwise antislice, and Figure 9-5b shows a counterclockwise antislice.

One final bit of notation will help in explaining the details of the solution. In addition to the six faces, it is important to distinguish the three middle layers of the cube. We will denote these as the X layer, Y layer, and Z layer. The X layer contains the white, green, yellow, and blue center facelets; the Y layer contains the blue, orange, green, and red center facelets; and the Z layer contains the white, orange, yellow, and red center facelets. Figure 9-6 highlights the three layers.

---

[*]David Singmaster, *Notes on Rubik's "Magic Cube,"* Penguin Books, 1981.

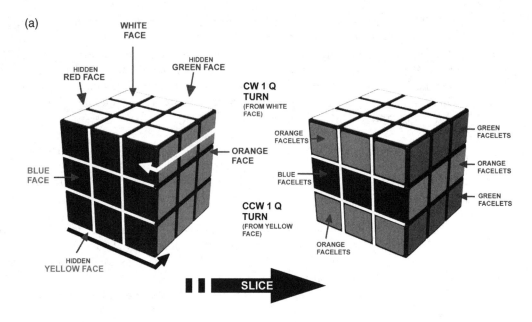

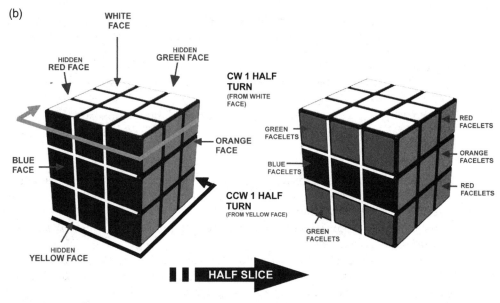

FIGURE 9-4 The effect of two slice moves on a solved cube: (*a*) a slice move; (*b*) an antislice or squared-slice move.

(a)

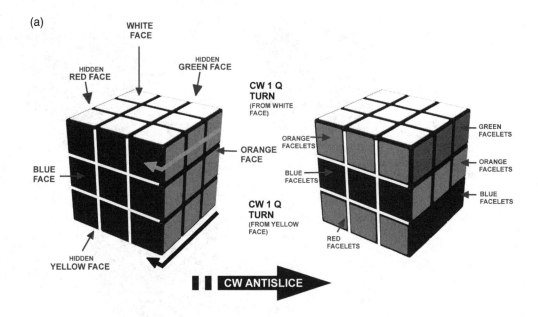

(b)

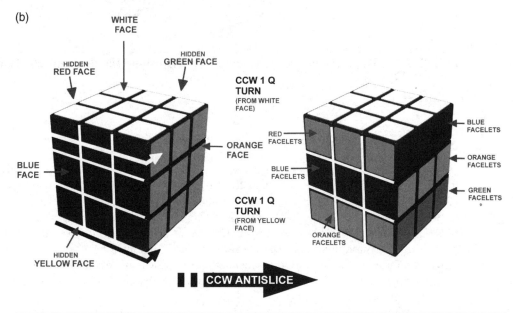

FIGURE 9-5 Contrasting antislice moves on a cube: (*a*) a clockwise antislice move; (*b*) a counterclockwise antislice move.

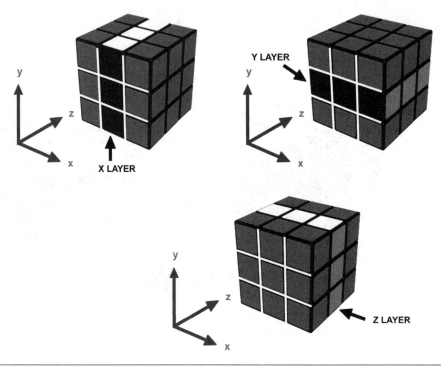

FIGURE 9-6 The three layers (X, Y, and Z) distinguishing the three middle layers of the Rubik's Cube.

Solving the Cube

A Rubik's Cube typically comes with a solution manual. The solution is relatively easy to follow, progressing in seven phases that solve the cube in a layer-by-layer manner. However, the solution is not very efficient because it generates a long sequence of twists—often in excess of 100 twists—and frequently moves an already placed piece out of position and later restoring it. A better solution was devised by Morwen Thistlewaite, a mathematician. Using group theory, Thistlewaite reasoned that a more efficient solution could be found by restricting the moves necessary to solve the cube. His approach was to use three intermediate solutions, each working toward the fourth and final solution of the cube.

A scrambled cube can be solved by applying a sequence of twists coming from the set $T_0 = \{B, G, O, R, Y, W\}$ and its derivatives, such as G' or W^2—a total of 33 different twists or 18 different twists ignoring slices and antislices. The robot will apply the minimum number of twists from this set to put the cube in the first intermediate position. A cube that has reached the first intermediate solution can be solved with twists from the set $T_1 = \{B, G, O^2, R^2, Y, W\}$ and their derivatives, for a total of 25 different twists, 14 without slices or antislices.

The difference between these two sets of twists has an interesting ramification. As an example, consider a solved cube and specifically the white-blue edge. After the twist sequence B–O–W, the edge is back in its original position, but its colors are flipped (other cubies are out of place, but that's okay because this example just concerns the white-blue edge). In fact, the only way for an edge to have its colors flipped in this manner is to perform quarter-turn twists along all three axes of the cube. Note that the T_1 set of twists only allows the red and orange faces to be twisted by half turns. This prevents edges from flipping because a quarter turn of either the red or orange faces is necessary to flip an edge. Therefore, once the cube has reached the first intermediate solution, edges can no longer be flipped.

This inability to flip an edge is the only difference between the T_0 and T_1 sets. Therefore, the first part of the solution focuses exclusively on removing flipped edges from the cube.

Once the first intermediate solution is reached, twists are performed to move the cube to the second intermediate solution. Once this solution is reached, the cube can be solved by performing twists from the set $T_2 = \{B^2, G^2, O^2, R^2, Y, W\}$, which consists of 17 different twists, 10 twists not including slices or antislices.

As with the first intermediate solution, a cube in the second intermediate solution has an interesting property: Facelets on the white and yellow faces can only move between the white and yellow faces. Therefore, the white face can only consist of white and yellow facelets, as does the yellow face. As a corollary, all the corner cubies will be properly oriented (white or yellow facelet on the white or yellow face). Another feature of this property is that the edge cubies from the Y layer (the middle layer of the cube containing the blue, orange, green, and red center facelets) are all in that layer but possibly in the wrong position. Twists from T_2 either swap adjacent edges in the Y layer or leave those edges alone; edges cannot move into or out of the Y layer.

The second part of the solution looks at two aspects of the cube:

- The *orientation* of the eight corner cubies, that is, which facelet of the corner is on the white or yellow face. When the cube is in the second intermediate solution, this facelet will be either white or yellow.
- The *position* of the four edge cubies that have neither white nor yellow facelets. As discussed earlier, these four cubies will be in their proper layer (the Y layer) once the second intermediate solution is reached.

Twists are then applied to the cube to move it toward the third intermediate solution; once there, the cube can be solved using twists from the set $T_3 = \{B^2, G^2, O^2, R^2, Y^2, W^2\}$, having nine different twists or six discounting slices. In this configuration, the corners are all in their proper *orbit* and orientation, and each edge is in its proper layer but possibly in the wrong position within the layer. A corner's orbit is the set of positions a corner cubie can occupy by performing only half twists of the cube; its orbit includes its home position. Note that a corner's orbit contains half the corner positions, arranged so that each face has two of the orbit positions, in opposite corners.

The third part of the solution focuses on two aspects of the cube:

- The position of each of the corner edges. When the cube is in the third intermediate solution, the corners are in the correct orbit with the correct orientation.
- The two layers (the X and Z layers) containing the eight white and yellow edge cubies. Edges in the wrong layer are placed in the correct layer, but possibly not the correct final position.

Once the cube is in the third intermediate solution, it is solved by placing the cubies in their correct positions using moves from the T_3 set.

Why This Approach?

Thistlewaite's approach has two main advantages that lend themselves to a computer-generated solution:

- The sequence of twists generated to solve a cube is far shorter than the layer-by-layer method.
- Each step of the solution can be precomputed and stored in lookup tables.

The solutions generated in this project contain at most 42 twists, with an average of slightly more than 30 twists. Although this is more than the theoretical minimum of 20—the so-called God's number—it is far better than the layer-by-layer method. There are algorithms that provide shorter solutions; for example, Kociemba's algorithm, which is related to Thistlewaite's, generates shorter solutions, but the solutions cannot be generated by table lookup (one lookup table would have in excess of 19.5 billion entries).

The general algorithm to solve the cube is as follows:

```
while Cube is not in a (partially) solved position, do
   Compute an index which encodes the properties of interest for
   the current phase
   Consult the lookup table, using the index to get the twist to apply
   Apply the twist to the Cube
end while
```

The algorithm is divided into a preparatory phase and four main phases, each main phase using a different twist set. Each phase has its own lookup table and its own method for encoding the relevant properties of the cube. The lookup tables are constructed so that for each phase the sequence of twists that is generated from the start of the phase until the cube reaches a solved position is as short as possible.

The size of the lookup tables is determined by the number of possible configurations of the cubies that are used in each phase. Table 9-5 shows the lookup table size and number of possible cube configurations at the end of each phase.

TABLE 9-5   The Size of the Lookup Table and the Number Cube Configurations at the End of Each Phase

Initial Number of Positions: 43,252,003,274,489,856,000		
Phase	**Entries in Lookup Table**	**Possible Configurations after Phase**
1	2,048	21,119,142,223,872,000
2	1,082,565	19,508,428,800
3	2,822,400	6,912
4	6,912	1

Phases of the Solution

The solution can be broken down into five phases. Table 9-6 summarizes the phases of the solution.

TABLE 9-6   Summary of the Phases Approach Algorithm

Phase	Description
Phase 0: Preparation	The cube is scanned and converted into four lists of facelets, where the facelet colors are identified by a number.
Phase 1: Unflip edges	The edges are unflipped, allowing them to be properly oriented with the limited set of twists available in later phases.
Phase 2: Part 1: Orient the corners Part 2: Edge positioning	The red and orange faces may only be twisted in half turns. Here the corners are oriented so that no white or yellow facelets appear on the blue, orange, green, or red faces.
Phase 3: Part 1: Corner positioning Part 2: Edge positioning	Only the white and yellow faces can be twisted by a quarter turn; all others must be twisted by half turns. This phase places cubies so that each cube face consists of two colors from opposite sides.
Phase 4: Part 1: Corner positioning Part 2: Edge positioning	All twists are half turns. The cube is solved at the end of this phase.

Phase 0: Preparation   The cube is scanned using the Android device's camera and converted into a list of facelets. The facelet list is then converted into four lists:

- A list of corner positions; if $p_c[i] = j$, then corner j is in position i.

- A list of corner orientations, with the following values:
  - $o_c[i] = 0$—The corner is properly oriented (white or yellow facelet on the white or yellow face).
  - $o_c[i] = 1$—The corner must be twisted counterclockwise to properly orient it.
  - $o_c[i] = 2$—The corner must be twisted clockwise to properly orient it.
- A list of edge positions; if $p_e[i] = j$, then edge j is in position i.
- A list of edge orientations; if $o_e[i] = 1$, then edge i is flipped. That is, if only twists from the T_1 set are used, then the edge will be flipped in its home position. If $o_e[i] = 0$, then the edge is not flipped.

The cubies are identified by a number rather than by facelet colors for ease of processing. The corner cubies are numbered as listed in Table 9-7.

TABLE 9-7 Numbered Corner and Facelets Colors

Corner Number	Facelet Colors
0	White, red, green
1	White, green, orange
2	White, orange, blue
3	White, blue, red
4	Yellow, green, red
5	Yellow, orange, green
6	Yellow, blue, orange
7	Yellow, red, blue

The edge cubies are numbered as listed in Table 9-8.

TABLE 9-8 Numbered Edge Cubies and Facelets Colors

Edge Number	Facelet Colors
0	White, green
1	White, orange
2	White, blue
3	White, red
4	Green, red
5	Orange, green
6	Blue, orange
7	Red, blue
8	Yellow, green
9	Yellow, orange
10	Yellow, blue
11	Yellow, red

This numbering scheme applies both to the edge and corner cubies and to their positions. If edge i (or corner i) is in position i, then the cubie is in its proper position but may be oriented incorrectly.

Generating the corner-position, corner-orientation, and edge-position lists is straightforward. To determine edge positions, a list of colors for each edge is precomputed. The pair of facelets from each edge of the scanned cube is compared against this list. One entry in the list will match; this is the edge to store in the list. This process is repeated for the corners. Corner orientation is determined by the order in which the corner facelets match the color list.

Determining flipped edges is less straightforward; this project uses a technique employed by Ryan Heise in his Human Thistlewaite algorithm.[*] An edge is considered flipped if any of the following apply to it:

- The edge has a blue or green facelet on the white or yellow face.
- The edge has a white or yellow facelet on the blue or green face.
- The edge has a facelet on the blue or green face and its *other* facelet is blue or green.
- The edge has a facelet on the white or yellow face and its *other* facelet is white or yellow.

If none of these apply, then the edge is not flipped.

This process also can be performed by lookup. Each edge facelet has two colors that would mark its edge as flipped; these colors are stored in a list and compared against each edge facelet.

Phase 1: Unflip the Edges   There are 12 edges, each of which may be flipped or unflipped. However, flips must occur in pairs, so there are always an even number of edges flipped. There is no sequence of twists to flip just a single edge. So 11 edges can be flipped or unflipped independently, and the twelfth edge is flipped or unflipped to make the total number of flipped edges even. Therefore, there are $2^{11} = 2,048$ possible configurations. The orientation of an edge changes only when the edge is on the red (or orange) face, and the red (or orange) face is twisted a quarter turn clockwise or counterclockwise. In those twists, the orientation of all edges on the red (or orange) face switches; unflipped edges become flipped, and flipped edges become unflipped. Half turns of the red or orange faces or any twist of the other faces do not change the orientation of any edge.

Each configuration is encoded by using a simple binary conversion. The edges are numbered 0 through 11. An encoding is an 11-bit value where $b_i = 1$ if edge i is flipped and $b_i = 0$ if the edge is unflipped; note that bit 11 is not used in the encoding because it is only a parity bit. The formula is

$$E = o_e[10] \times 2^{10} + o_e[9] \times 2^9 + o_e[8] \times 2^8 + \ldots + o_e[1] \times 2^1 + o_e[0] \times 2^0$$

---

[*]http://www.ryanheise.com/cube/human_thistlewaite_algorithm.html.

This 11-bit number is the index of the lookup table. An encoding of 0 is the goal because this indicates that all edges are unflipped.

Phase 2: Part 1: Orient the Corners; Part 1: Edge Positioning   There are eight corners, each of which can be in one of three orientations. Seven of the corners can be oriented independently, whereas the eighth corner acts as a parity. Using the encoding of the corner-orientation list, the sum of the values in the list must be a multiple of 3. The corners are oriented in one or more of the following patterns:

- Individual corners may be properly oriented.
- Pairs of corners can have opposing orientations, one corner twisted clockwise and one counterclockwise.
- Three corners can all be twisted in the same direction.

Corner orientation is encoded by treating the orientation list as a base-3 number, ignoring the most significant digit (the corner in position 7) because it is parity. The formula is

$$c = o_c[6] \times 3^6 + o_c[5] \times 3^5 + o_c[4] \times 3^4 + \ldots + o_c[1] \times 3^1 + o_c[0] \times 3^0$$

One of the goals in phase 2 is to properly orient the corners, that is, twist all the corners so that their white or yellow facelets are on either the white or yellow faces. A corner's white facelet may be on the yellow face or vice versa; this will be corrected in the last two phases.

The other goal of phase 2 is to position the edges so that a white or yellow facelet is on either the white or yellow face. As with corners, the edge's white facelet may be on the yellow face or vice versa; this will be corrected in phases 3 and 4. From a programming perspective, it is easier to consider the four edges without white or yellow facelets. In phase 2, these four edges are moved to the Y layer, but they can be in any permutation within that layer. Encoding the edge positions is done by using a combinatorial counting system described in Knuth.[*] Suppose that the four nonwhite/nonyellow edges are in positions p_1 through p_4, with $p_1 > p_2 > p_3 > p_4$. Then this combination can be encoded as

$$e = \binom{p_1}{4} + \binom{p_2}{3} + \binom{p_3}{2} + \binom{p_4}{1}, \text{ where } \binom{n}{k} = \frac{n!}{n!(n-k)!}$$

The encoded number has 495 possible values, corresponding to the number of ways to choose 4 items from a set of 12, with a range of $0 \leq e < 495$. A combined encoding that captures corner orientation and edge positions can be computed using $E = c \times 495 + e$. There are thus $495 \times 2{,}187 = 1{,}082{,}565$ configurations that are considered in phase 2.

---

[*]Donald E. Knuth, *The Art of Computer Programming, Volume 4, Fascicle 3: Generating All Combinations and Partitions*, Addison-Wesley Professional, 2005.

An encoding of $69 = 0 \times 495 + 69$ is the goal; in this position, the corners are all properly oriented ($c = 0$), and the four nonwhite/nonyellow edges are in the Y layer, positions 7, 6, 5, and 4:

$$e = \binom{7}{4} + \binom{6}{3} + \binom{5}{2} + \binom{4}{1} = 69$$

Phase 3: Part 1: Corner Positioning; Part 2: Edge Positioning   In this phase, the corners are put into their proper orbits, and the edges are moved to their proper layer (note that the four nonwhite/nonyellow edges are already in their proper layer at the start of phase 3). There are $8! = 40,320$ ways to place the corners. The corner permutations can be encoded using the permutation-counting method described in Knuth.[*]

Given the corner-position list p_c, define $C[i]$ to be the number of elements in p_c such that:

$$j > i \text{ and } p_c[j] < p_c[i]$$

Then the corner positions can be encoded as

$$c = 7! \times C[0] + 6! \times C[1] + \ldots + k! \times C[7 - k] + \ldots + 1! \times C[6]$$

(note that $C[7]$ will always be 0). There are $24^2 = 576$ permutations of the corners within the two orbits. However, only 96 of these allow the cube to be solved with twists from the T_3 set. The encodings for this set of valid permutations can be found in **phase3gen.cc**, the program that creates the lookup table for phase 3. Any of these 96 permutations is considered a solution for phase 3.

The other goal of phase 3 is to place the edges in their proper layer. The edges that belong in the Y layer are already there, so this phase will exchange edges between the X and Z layers as appropriate. Because edges are exchanged, there must always be an equal number of incorrect edges in these layers.

There can be any number of incorrect edges in the X layer, and the location of the incorrect edges in the X layer does not affect the location of incorrect edges in the Z layer. This means that there are

$$e = \binom{4}{0}^2 + \binom{4}{1}^2 + \binom{4}{2}^2 + \binom{4}{3}^2 + \binom{4}{4}^2 = 70$$

ways in which the X and Z layer edges can be in the wrong layer. An encoding based on this formula and the combinatorial counting method from phase 2 is used to encode the X and Z layer edges. The encoding for this phase is then $E = c \times 70 + e$, where e is the encoding of the edges in the X and Z layers. Phase 3 is completed when $E = 70 \times c_s$, where c_s is any of the 96 valid corner permutations. At this point, the

---

[*]Donald E. Knuth, *The Art of Computer Programming, Volume 4, Fascicle 2: Generating All Tuples and Permutations*, Addison-Wesley Professional, 2005.

corners are in their orbits and can be solved using only twists from T_3, and the edges are in their proper layer.

Phase 4: Part 2: Corner Positioning; Part 3: Edge Positioning   The final phase puts the cubies into their correct positions. At the beginning of the phase, the edges are in their correct layer but possibly out of order. Two of the layers can be in any of the 4! = 24 permutations of their four edges, whereas the third layer can only be in one of 12 permutations depending on the exact permutation of the other two layers. This gives rise to 6,912 possible positions. The positions can be encoded using the same permutation-counting technique from the preceding section. If p_x, p_y, and p_z are the permutation numbers for the X, Y, and Z layers, respectively, then the cube can be encoded by $e = p_x \times 24^2 + p_y \times 24 + p_z$. The corner cubies can be in one of 96 permutations. The cubies in corners 0, 2, 5, and 7 (orbit 0) can be in any permutation, whereas the cubies in corners 1, 3, 4, and 6 (orbit 1) can be in one of four permutations. The orbit 0 cubies are encoded using the permutation-counting method described in phase 3, whereas the orbit 1 cubies are identified by which cubie is in corner position 1. The corner permutation is encoded using $c = p_0 \times 4 + p_1$, where p_0 and p_1 are the permutation numbers for orbits 0 and 1, respectively. The overall encoding is given by $E = c \times 6912 + e$, with $E = 0$ representing the solved cube.

Cube Solver Design

The cube solver will unscramble a Rubik's Cube using the four-phase algorithm just outlined. The hardware consists of six basic sections:

- A twister consisting of a stationary servo with an attached holder on which the cube sits
- A second twister consisting of a sliding servo with an attached holder that sits on the cube
- A slider that pushes the sliding servo
- A retractable gripper that can grip the center layer of the cube
- An alignment module that ensures that the cube faces are properly twisted
- A bracket that holds the Android device

The cube is twisted by gripping the cube, which holds one of the X, Y, or Z layers stationary, and then spinning one or both of the holders, twisting the top and/or bottom face of the cube. To twist a different face, the cube is flipped by sliding the top servo, partially flipping the cube. The gripper is then extended to push the cube back onto the lower tray, completing the flip.

The robot contains five servos. However, the standard NXT system can only control three. The robot therefore uses two NXT computers. One NXT unit controls the gripper and slider because these need to work cooperatively to flip the cube. The other NXT unit controls the twisters and aligner. Table 9-10 shows the capability matrix for the cube solver robot, which lists the main hardware.

TABLE 9-10   The Capability Matrix for the Cube Solver Robot

Team B Members	Microcontroller	End Effectors	Mobility Type	Display	Sensors/ Motors	I/O Port	Ranges	Other Capabilities
Twisty (B1R2)	32-bit ARM	Upper: Twister (rotating platform, oriented upside-down)	N/A	100- × 64-pixel liquid-crystal display (LCD) screen	Upper twister	A	0–360 degrees	N/A
		Lower: Twister (rotating platform)			Lower twister	B	0–360 degrees	
		Aligner: Two fingers moving in parallel			Aligner	C	–2,300–0 degrees	
Flippy (B1R1)	32-bit ARM	Flipper	N/A	100- × 64-pixel LCD screen	Flipper (linear actuator)	A	–120–0 degrees	N/A
		Gripper			Gripper (linear actuator)	B	–45–0 degrees	
Android	32-bit dual-core ARM		N/A	800- × 400-pixel WVGA	Camera	N/A	N/A	N/A

Design Issues

The main design difficulty involves the dimensions of Rubik's Cubes and LEGOs. A standard Rubik's Cube is 57 mm wide. LEGO bricks and beams are 7.8 mm wide, and LEGO studs and Technic holes are 8 mm on center. Thus any combination of LEGO beams is either 1 mm too small or 7 mm too large. The twisters each use a 9-stud × 9-stud holder; the beams in the center 7-stud × 7-stud square are oriented so that the holes are horizontal; this minimizes friction with a flipping cube. The outer edges have their holes aligned vertically. Connector pins are inserted along the edge to hold the cube as it spins. Since the pins are not 8 mm wide, this allows the cube to be held securely enough to twist a face while providing enough space for the cube to be flipped. Care must be taken, however, to account for the "wiggle room" that the Cube has while sitting in the holder. When making a twist, the holder must be twisted slightly more than 90 degrees (or 180 degrees for a half turn) in order to get the cube face aligned properly, and then the holder must be twisted slightly in the opposite direction to align the holder.

The gripper consists of two stationary arms with an inside distance of 56 mm. Although the arms are set 1 mm too close together, there is enough give in the arms to allow them to slip around the cube, providing a snug fit. Angled lift arms are used at the end of both arms, allowing the gripper to fit around a cube that is slightly out of alignment.

Cube Solver Hardware: The Frame

The frame consists of two parts. The right section holds the gripper and alignment mechanisms, whereas the left section houses both twisters and the flipper mechanism. The two sides slide together and are connected with eight pins.

Lower and Upper Twister: Twisting Left and Right

Figure 9-7 shows the lower twister mechanism. The lower twister has a fairly basic design. The servo rests on two crossbars at the base of the model. A third crossbar consisting of a five-unit axle and two three-unit axles at the rear of the servo adds stability; without it, the servo and twister base have a tendency to sag to one side

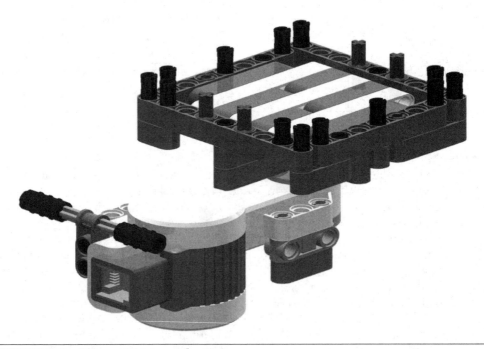

FIGURE 9-7   Lower twister mechanism.

because the weight of the mechanism (with the cube sitting on it) is slightly off-center. The twister base is built to have as smooth a surface as possible. During the flip process, an edge of the cube must slide across the twister base. The smooth surface minimizes the chance that the edge will be caught on the edge of one of the LEGO pieces and jam the system.

The servo rotates 90 degrees for a clockwise quarter turn and 180 degrees for a half turn. For counterclockwise quarter turns, the servo rotates 70 degrees prior to alignment. These choices allow the alignment mechanism to nudge the bottom face of the cube into alignment with the center layer. After alignment, the servo rotates an additional 20 degrees to align the twister base.

As its name implies, the upper twister turns the top face of the cube. It is also part of the flipper mechanism, pushing the cube partially onto its side. Since the cube does not rest on the twister, the twister's face does not have to be as smooth as the lower twister's face and thus is built with fewer parts. In particular, the upper twister has four exposed connector pin ends that could cause the cube to stick if they were on the lower twister.

The upper twister will rotate the servo 90 degrees for a quarter turn, 180 degrees for a half turn, and 110 degrees for a counterclockwise quarter turn. Thus any error in alignment will favor the top face being twisted counterclockwise relative to the center layer, regardless of which twist is performed. This allows the alignment mechanism to correct the alignment error. A clockwise misalignment would not allow the alignment fingers to nudge the layers back into place. After alignment, the servo turns 20 degrees clockwise after a clockwise quarter turn to realign the twister.

Gripper and Aligner: Hold On, Straighten Up, and Twist Right

Figure 9-8 shows the gripper mechanism. The gripper mechanism serves a dual purpose. First, it holds the center layer of the cube stationary while the twister mechanisms turn the top and bottom faces as necessary. Second, it works with the flipper mechanism to flip the cube.

The gripper is a fairly simple mechanism. It consists of a slide, servo, and mounting frame. The slide has two fingers that fit around the center layer of the cube; the fingers bend away from the cube, allowing them to correct minor errors in the cube's alignment. A small bumper at the back end of the fingers prevents the gripper from pushing too far forward, which would lodge the cube too tightly in the mechanism. There is a 1-mm difference between the width of the cube and the distance between the fingers. The LEGO pieces have enough play in them to accommodate the difference in the outer half of the fingers, but closer to the crossbar there is less ability to spread apart; the bumper keeps the cube at a safe distance from the crossbar. Two exposed long connector pins serve as a stop for the gripper as it retracts. This keeps the gripper from retracting too far and also serves as a point of origin for measuring how far the servo needs to rotate. Without such a point, gear lash and variance in how far the gripper advances eventually would move the gripper out of alignment. The servo has a range of motion of about 45 degrees. At 0 degrees, the gripper is retracted to the stop pins. At 45 degrees, the gripper is at the cube, pushing slightly against it.

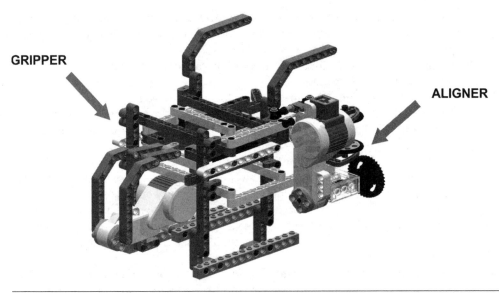

FIGURE 9-8 Gripper mechanism with the alignment module attached.

The mounting frame is designed to hold the slide at the proper elevation so that it grips the center layer of the cube without hindering the top or bottom layers. It also prevents the slide from twisting; as the top and bottom cube faces are twisted, they apply torque to the gripper. Without the side guides to hold the slide in place, certain twists would not complete a quarter turn because the center layer also would turn a small but significant amount. The gripper frame also holds the alignment module, described next.

The flipper mechanism is particularly sensitive to the alignment of the cube. If the top or bottom face is not aligned properly with the center layer, the cube will not turn properly during the flip process; this typically results in the cube becoming dislodged and jamming the system. To prevent this situation, the alignment module helps to make the top and bottom layers line up with the center layer.

The module consists of two "fingers"—a lift arm, wheel, and tire—that swing out from their home position and nudge the layers into a better alignment. The fingers then swing back to their home position, out of the way of the cube and gripper mechanism.

Flipper: Flip It Well and Good!

Figure 9-9 shows the top motor assembly and the flipper and twister mechanisms. The flipper mechanism works in conjunction with the gripper to flip the cube so that its top and bottom faces flip to either side, allowing all six sides to be twisted. The flipper consists of a sliding bracket that holds the upper twister and its servo, two rails along

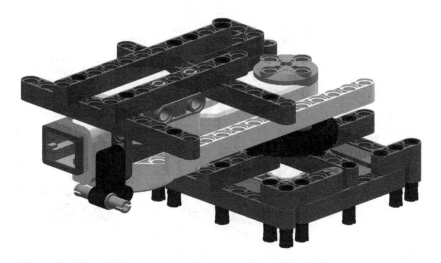

FIGURE 9-9   The top motor assembly showing the flipper and upper twister.

which the flipper slides, and a servo that drives the flipper. The bracket consists of two beams that rest on either side rail to keep the mechanism from falling, two beams that run along either side of the twister servo to keep the mechanism aligned, and a few other pieces that attach the twister servo to the bracket. The side rails form part of the cube solver frame. A crossbar at the rear of the rails prevents the two sides from spreading apart and allowing the upper twister to fall or wedge itself between the rails.

The flipper servo is mounted on two beams at the far left edge of the cube solver. Taking advantage of a 3–4–5 right-triangle alignment of holes on the servo allows it to be mounted on two rails rather than three or four while still maintaining a sturdy, rigid structure. A two-part linkage connects the servo to the rear of the upper twister servo. The mechanism allows the upper twister to slide forward, forcing the cube to partially tilt over. The gripper then extends, forcing the cube back over the lower twister. The upper twister then retracts, and the gripper extends and retracts, nudging the cube back into the lower twister and completing the flip.

Android Bracket and NXT Stand

Figure 9-10*a* shows the NXT stand as separate from the rest of the cube solver model. It isn't necessary for it to be attached, and there is no convenient attachment point. Cables connecting the NXT units to the servos enter the cube solver frame at the left end of the model, mostly underneath the flipper and lower twister servos.

Figure 9-10*b* is the bracket that holds the Android-based smartphone in a fixed place relative to the cube, allowing it to take pictures of the cube faces to determine facelet colors. It is tilted slightly to allow better visibility for the user.

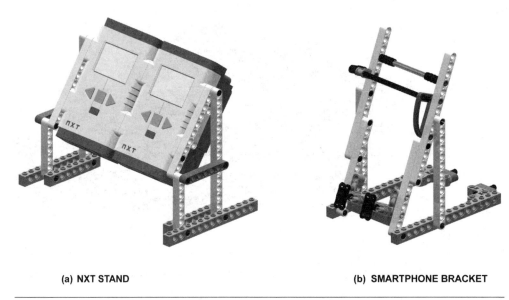

(a) NXT STAND **(b) SMARTPHONE BRACKET**

FIGURE 9-10 (*a*) The NXT stand and (*b*) the smartphone bracket.

Cube Solver Software

The software for the cube solver consists of 12 modules, all written in Java. There are three reasons for this choice of language over a LabVIEW/NXT-G environment:

- Java is the language of choice for Android development.
- leJOS is a freely available, robust Java environment for the NXT.
- The heavy use of math and lookup tables lends itself more to a conventional programming language than the NXT-G environment.

Two of the modules—**Twisty** and **Flippy**—are loaded onto their respective NXT units. The nine modules form the cube solver app for the Android.

Twisty Module: Turning This Way and That Way

Both the twisters are controlled by the **Twisty** NXT. The **main()** function of **Twisty** consists of three sections: initialization, communication loop, and shutdown. The initialization section establishes motor speed, draws the initial message on the LCD screen, and waits for the Android to initiate communication. The command loop is a basic **if-else-if** chain inside an infinite loop. Each **if** statement processes one of the commands. The loop looks like this:

```
while (true) {
        try {
            cmd = dis.readByte();
```

```
            check = dis.readByte();
            check++;

            if (cmd == Command.PING.getCode())
                drawCmd("PING");
            ...
            } else if (cmd == Command.TWIST_TOP_CW.getCode()) {
                drawCmd("Top");
                twistTopCW();
            } else if (cmd == Command.TWIST_TOP_CCW.getCode()) {
                drawCmd("Top'");
                twistTopCCW();
            ...
            } else if (cmd == Command.ALIGN_ON.getCode()) {
                drawCmd("YesAlign");
                shouldAlign = true;
            } else {
                // all other commands are an error
                drawCmd("Error "+cmd);
                rc = 3;
            }

        dos.writeByte(rc);
        dos.writeByte(check);
        dos.flush();

        if (cmd == Command.SHUTDOWN.getCode())
            break;
    } catch (IOException e) {
        e.printStackTrace();
    }
  }
}
```

The shutdown section closes the communication channel to the Android, shuts off the motors, and terminates the program.

The twist commands are performed by 11 helper functions, which are listed in Table 9-11 with a short description of each. These rotate the servos to their proper position to effect the twist. One such function is listed here; the other functions are identical in structure.

```
private static void sliceTop() {
    int a = Motor.A.getTachoCount();
    int b = Motor.B.getTachoCount();
    Motor.A.rotateTo(a+95, true);
    Motor.B.rotateTo(b-95, false);
    Motor.A.rotateTo(a+90,true);
```

TABLE 9-11   The 11 Twist Command Helper Functions

Twist Command Helper Functions	Description
sliceHalf()	Twists top and bottom faces 180 degrees
sliceTop()	Twists top face 90 degrees clockwise and twists bottom face 90 degrees counterclockwise
sliceBottom()	Twists bottom face 90 degrees clockwise and twists top face 90 degrees counterclockwise
twistTopHalf()	Twists top face 180 degrees
twistTopCW()	Twists top face 90 degrees clockwise
twistTopCCW()	Twists top face 90 degrees counterclockwise
twistTopCCW()	Twists bottom face 180 degrees
twistBottomCW()	Twists bottom face 90 degrees clockwise
twistBottomCCW()	Twists bottom face 90 degrees counterclockwise
antisliceCW()	Twists top and bottom faces 90 degrees clockwise
antisliceCCW()	Twists top and bottom faces 90 degrees counterclockwise

```
    Motor.B.rotateTo(b-90,false);
    if (shouldAlign)
        align();
}
```

The Flippy Module: Gripping and Flipping

The **Flippy** module controls the gripper and flipper servos. Its **main()** function is almost identical to the **Twisty** module; the only difference is in the set of commands processed by the **if-else-if** chain. Three helper functions process the three movement commands. Two commands grip and release the cube, whereas the third command flips the cube. These helper functions are listed with a short description in Table 9-12. To grip the cube, the gripper servo is rotated to a position of –45 degrees; to release the cube, the servo is rotated to a position of 0 degrees. Both positions are relative to the starting position, which has the gripper fully retracted. Flipping the cube begins by rotating the flipper servo to a position of –120 degrees relative to its starting position, which slides the upper twister forward, partially flipping the cube. The cube will be pushed slightly into the gripper mechanism, holding it in place.

The servo is then rotated back to its starting position, causing the upper twister to slide over the cube into its home position. The cube is held in its partially flipped position by the gripper. Next, the gripper extends, forcing the cube back. The gripper then retracts quickly, allowing the cube to fall. At this point, the cube has one edge

TABLE 9-12   The Flippy Helper Functions

Flippy Helper Functions	Description
flip()	Flips the cube
grip()	Extends the gripper around the cube
release()	Retracts the gripper from the cube

resting on the lower twister, whereas the other edge rests on the pins used to grip the cube while twisting. The gripper extends one more time to push the cube back so that it is fully over the base of the lower twister and then retracts to allow the cube to drop onto the base.

CubeSolverActivity Module: Initiating the Android App

This is the main module for the Android app. Figure 9-11 shows the class relationship diagram for CubeSolverActivity. It sets up the user interface, creates and initializes the Communicator and Solver objects, and provides several callback functions to assist other modules (mainly the Visual and Solver modules). The onCreate() method creates the user interface and sets up the callback functions

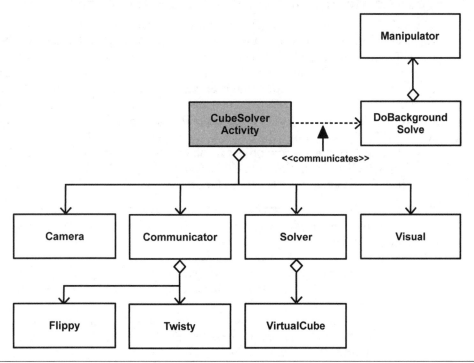

FIGURE 9-11   The class relationship diagram of class CubeSolverActivity.

for the user interface buttons. The `surfaceCreated()` and `surfaceChanged()` callbacks, which are invoked as the `SurfaceView` object is set up, are also involved in this process. They create and initialize the Android's camera.

Two private classes are employed: `doBackgroundInit()` and `doBackgroundSolve()`. These allow the Android to process time-consuming tasks while not tying up the user-interface thread. The `doBackgroundInit()` class creates and initializes the `Communicator` and `Solver` objects, both of which are long tasks from the Android's perspective. The `Communicator` takes time to establish the Bluetooth links to `Twisty` and `Flippy`, whereas the `Solver` must load the lookup tables for each of the four phases.

The `doBackgroundSolve()` class performs the following steps:

- Send the facelet information to the `Solver` object and obtain a solution.
- Use the `Manipulator` object to apply the solution to the cube.

The user-interface thread is used to obtain the facelet images. The camera is started in the `takePicture()` method; when the picture is finished, the `onPictureTaken()` callback is invoked, which invokes the `Visual` module's `getVisualData()` method, and then calls `takePicture()` to obtain the image of the next cube face.

Solver Module and Helpers: Solving the Virtual Cube

The `Solver` module actually consists of six Java classes: one for a virtual representation of the cube, one for each of the four phases, and one for the `Solver` itself. The `Solver` is given a 54-character string, representing the colors on each facelet on the cube. It then creates a virtual representation of the cube consisting of the edge- and corner-position and orientation lists, as described in the solution section. The `Solver` then solves the virtual cube, creating the list of twists necessary to solve it. The twist list is then returned to the `Manipulator`, which communicates with the NXT units to apply the twists to the physical cube. Figure 9-12 shows the pipeline transformation of the physical cube to its virtual representation, ultimately producing the list of commands used to solve the physical cube.

The basic `Solver` algorithm is as follows:

```
Create the four lists — corner position, corner orientation,
edge position, edge orientation.
For phase = 1 to 4, do the following:
    Loop forever, doing the following:
        Determine Cube configuration, based on the current phase.
        If Cube is in solved configuration for the current phase, stop
        the loop.
        Consult current phase's lookup table for the twist to apply,
        using the cube configuration as the index.
        Append the twist to the twist list.
        Apply the twist to the virtual Cube.
```

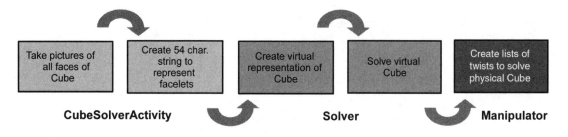

FIGURE 9-12   The pipeline transformation of the virtual cube to the list of twist commands and the modules that perform each transformation.

```
              Append end-of-phase marker to the twist list.
Return the twist list to the manipulator.
```

Details of each phase are given in the "Solving the Cube" section above. If the color string does not represent a valid cube (e.g., the Android incorrectly identified one or more facelet colors), the `Solver` will return a zero-length solution, indicating an error to the `Manipulator`. Note that an already solved cube will have a solution consisting of four end-of-phase markers so that there is no confusion between a solved cube and an incorrect identification.

Each of the four phases is almost identical, consisting mainly of a constructor that loads the lookup table and a `solve()` function that takes a `VirtualCube` object and applies the phase's solution algorithm to it. The `solve()` function for phase 2 is shown below:

```java
public void solve(VirtualCube r) {
    int e,i,j,k;
    byte b;

    while (true) {

        // encode Cube configuration

        j = 4;
        i = 0;
        for (e=11;e>=0;e--) {
            k = r.getEdge(e);
            if (k >= 4 && k < 8) {
                i += choose(e,j);
                j--;
            }
        }
        for (j=6;j>=0;j--)
            i = i * 3 + r.getCornerRotation(j);
```

```
        b = lookupTable[i];

        //
        // -1 is the stop code.
        //

        if (b == -1)
            break;

        r.applyTwist(b);

        solutionList[solutionList[0]++] = b;
    }

    solutionList[solutionList[0]++] = 99;
}
```

The **solve()** methods for the other phases are similar, with only the encoding instructions changing.

Visual Module: Identifying the Color of the Facelets

The visual data-acquisition section of the software consists of one primary function, **getVisualData()**, along with a few helper functions. The function takes a snapshot of one cube face, building up a string of letters corresponding to the facelets on the cube. The **onPictureTaken()** callback does the bulk of the work. As its name implies, it is called when the camera has finished obtaining its image. It identifies the color of each facelet using the following basic algorithm:

1. Scan an 11-by-11 square of pixels, averaging the RGB components
   separately.
2. Convert the average RGB value to an HSV value.
3. Use the hue and saturation levels to determine the color.

Generally, the pixel squares used in each of the samples are near the center of the facelet. However, the squares for the top-layer facelets are closer to the center layer to avoid sampling the connector pins, which are black. The top-center facelet sample square is in the lower-right corner of the facelet. This helps to avoid a problem with washout because the Android flash has a tendency to make the facelet too bright to discern the color if sampled in the center. Finally, the center facelet is sampled in the lower-right corner of the facelet. This avoids the "Rubik's Cube" logo on the center white facelet.

Using hue-saturation-value (HSV) data does a better job of discerning facelet colors than does red-green-blue (RGB) data. While RGB data are always dependent on the light level, HSV values are less so; within a reasonable range of lighting levels, differences in light level are mainly reflected in changes to the V component of an

HSV value. Aside from white, the facelet colors each occupy a fairly narrow band across the spectrum, translating into a narrow band of H values for each. These are easily discerned by an **if-else-if** chain. Typically, white is identified by a very low saturation level because saturation is a measure of color intensity. The **getVisualData()** method works in conjunction with the **takePicture()** method in the main **CubeSolverActivity** object. The **takePicture()** method controls the timing and sequencing of the pictures, rotating the cube between pictures and forcing the repeated **getVisualData()- onPictureTaken()** calls to pause between pictures until the camera is reset and the cube rotated for the next picture. After the sixth picture is processed, **takePicture()** starts the **Solver**.

Manipulator Module: Executing the Solution

The **Manipulator** is the section of software that manages execution of the solution. The **Manipulator** is given each twist in the solution list, one at a time. For each twist, it generates the following command sequence:

1. Orient the Cube by performing a slice move with the gripper retracted;
2. Flip the Cube;
3. Perform a twist, slice or antislice with the gripper extended.

The first two commands may not be necessary. If the desired cube face is already facing up or down, then the cube is already in the proper position for twisting, and only the third step is necessary. If the face to be twisted is facing right or left (as seen by the Android), then the cube is ready to be flipped, and the first step may be omitted. If the face to be twisted is facing front or back, then all three steps must be performed. In order to determine the orientation and flip for each twist, the **Manipulator** keeps track of the cube's physical orientation. This is separate from the **Solver** module's virtual orientation, which always keeps the white face up and blue face in front. The **Manipulator** uses a small array of characters to remember the center facelet of each side, rotating the values within the array to match the twists and flips the **Manipulator** performs. Note that performing one of the twists, slices, or antislices to turn individual cube faces does not alter the cube's orientation; the centers remain in their same positions after such a twist.

The **Manipulator** object works in conjunction with the **doBackgroundSolve()** method in the **CubeSolverActivity** object. This allows a two-threaded solution process, with one thread managing the user interface (primarily the information panel and "Stop!" button) and one thread managing the solution process. The background solver is launched from the **takePicture()** method after all six cube faces have been processed. It begins by sending the color information to the **Solver** object to obtain a solution. This is quite rapid because the algorithm for finding a solution is very simple, encoding the cube's current permutation notwithstanding. Once the solution sequence is generated, it is processed one at a time, displaying the twist on the information panel and then using the **Manipulator** to apply the twist to the cube. As mentioned in the **Solver** section, the **Solver** returns a zero-length solution to indicate an error in solving. The **Manipulator** will restart the facelet identification when the **Solver** indicates an error.

Communicator Module: Establishing Bluetooth Communication with Twisty and Flippy

The final module in the cube solver app is the `Communicator`. Figure 9-13 shows the class relationship diagram `Communicator`. The `Communicator` consists of two main methods: `connect()` and `sendCommand()`. The `connect()` method initiates Bluetooth communication with `Twisty` and `Flippy` in a four-step method. First, the Android's Bluetooth transceiver is activated:

```
if (!localAdapter.isEnabled()) {
        localAdapter.enable();

        for (int i=0;i<10;i++) {
                try {
                    Thread.sleep(1000);
                } catch (InterruptedException e) {
                    e.printStackTrace();
                }
                publishProgress(".");
            if (localAdapter.isEnabled())
                break;
        }
    }

if (localAdapter.isEnabled())
    publishProgress("<br/>Enabled.<br/>");
else {
    publishProgress("<br/>Error: Cannot enable Bluetooth.<br/>");
    return null;
}
```

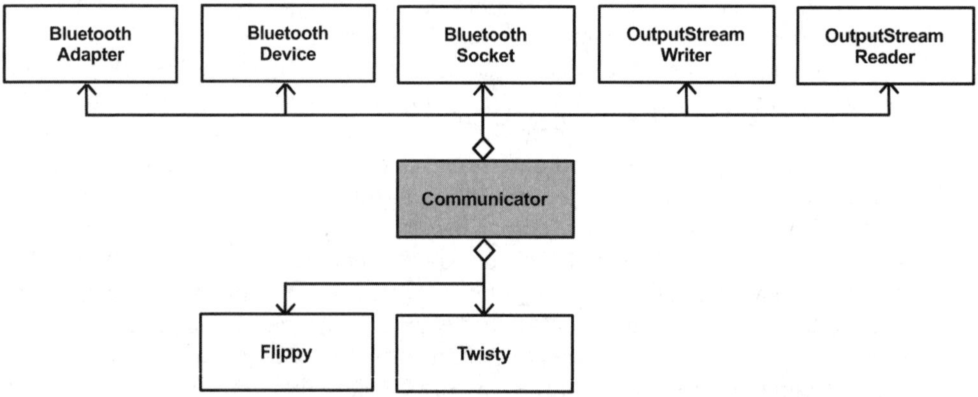

FIGURE 9-13   The class relationship diagram of class `Communicator`.

Next, two sockets are created, one for each of the NXT units:

```
flippyDev = localAdapter.getRemoteDevice(flippyMAC);

try {
    flippySocket = flippyDev.createRfcommSocketToServiceRecord(UUID
                    .fromString("00001101-0000-1000-8000-00805F9B34FB"));

    flippySocket.connect();

    flippyIn = new InputStreamReader(flippySocket.getInputStream());
    flippyOut = new OutputStreamWriter(flippySocket.getOutputStream());

} catch (IOException e) {
    Log.d("Bluetooth","Err: Device not found or cannot connect to flippy");
    publishProgress("<br/>Error: Cannot connect to flippy.<br/>");
    return null;
}
```

Finally, a `PING` command is sent to both NXT units to verify that communication is established. If all goes well, the `Solver` object is then created and initialized in the `CubeSolverActivity` module's `doBackgroundInit()` method. If an error occurs, any established communication links are closed, and the app is terminated. Table 9-13 lists the `Visual`, `Manipulator`, and `Communicator` modules and their methods with brief descriptions of each.

Setting Up Programming

Each of the four phases in the solution process uses a lookup table to determine the shortest sequence of twists necessary to complete the phase. These are precomputed by four programs written in C++. Each program uses the same algorithm—a breadth-first search (BFS)—to populate its lookup table. Theoretically, a BFS is equivalent to Dijkstra's shortest-path algorithm when each edge has equal cost but is far easier to implement and has a faster run time. For this project, the vertices in the BFS graph represent cube permutations, and edges represent a twist converting one permutation into another. The solution states are initially loaded into the BFS queue, and shortest paths back to a solution are generated.

To save space, only the twist information is stored in the lookup tables. Although it would be faster to also store the resulting permutation after the twist, doing so would quadruple the size of the lookup tables. The additional time required to recode the permutation after each twist is negligible compared with the time required to load the lookup tables from files, so this tradeoff is acceptable.

There are only two key differences between the generator programs. First, the algorithm for encoding a cube permutation is different for each phase, as discussed at

TABLE 9-13 The Methods for the `Visual`, `Manipulator`, and `Communicator` Modules with Brief Descriptions of Each

Helper Modules	Description
`Visual`	`public void clearColors()` clears the color string prior to scanning the cube.
	`public String getColors()` returns the 54-character string representing the cube facelets.
	`public void GetVisualData(final Context c)` scans one face of the cube, identifies facelet colors, and adds the colors to the color string.
`Manipulator`	`public void applyTwist(int t)` applies twist to the physical cube by sending appropriate commands to `Twisty` and `Flippy`.
	`private int orientFace(int f)` ensures that a face is either on the top or bottom of the physical cube; will rotate and/or flip the cube if necessary; returns 0 if the face is on top and 1 if it is on the bottom.
`Communicator`	`public boolean isConnected()` returns `true` if connections have been established with `Twisty` and `Flippy`.
	`public void connect()` establishes connection with the NXT bricks; sets the Boolean `connectStatus` based on the success of the connections.
	`public boolean sendCommand(NXT who, Command cmd)` sends command code `cmd` to the specified NXT. Returns `true` if the message was sent successfully.

the beginning of this chapter. Second, the allowable twists are different in each phase, with each successive phase having a smaller set of twists from which to select.

Running the Robot

To run the cube solver, the robot components have to be aligned properly:

- Both twisters should have their holders aligned with the frame.
- The gripper should be fully retracted.
- The flipper bracket should be retracted as far as possible (as far to the left as possible).
- The aligner fingers should be parallel to the back of the frame.

A scrambled Rubik's Cube is inserted. The cube rests on the lower twister, inside the connector pins. The upper twister rests on the cube with the connector pins around it. Note that the cube should be easy to twist (the servos may not have enough torque to twist a tight cube). Third, **Twisty** and **Flippy** programs are running. Fourth, the **CubeSolver** app is started on the Android. It takes a moment or two to initiate communication and load the lookup tables; the screen will be updated as it progresses. Once the app is ready, the "Solve!" button is pressed to begin the process. The screen will display its progress and the solution as each twist is applied. Figure 9-14 shows a screen capture of the solution display of the Android smart phone screen. Figure 9-15 is the cube solver in all its glory.

What to Do Next Time

The main issue with the cube solver is fault tolerance. The system can jam during a flip, causing the cube to dislodge (in fact, my daughter called an early prototype the "Cube Flinger" after watching it jam and eject the cube out of the robot entirely). There is currently no way to recover from a jam; the "Stop!" button will stop the robot, but this, of course, is not an automated process. The software is partly fault-tolerant and partly fault-aware. If communication links cannot be established, the system can detect this. However, since the robot team cannot function without all members

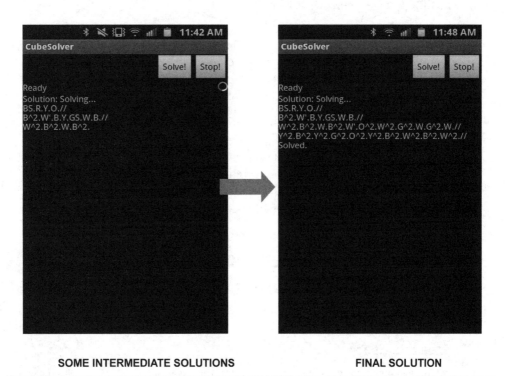

FIGURE 9-14   Screen capture of solution displayed on the Android smartphone.

FIGURE 9-15 The cube solver.

working, the system shuts down. Similarly, if communication fails while the team is working, the team will stop and shut down.

The **VirtualCube** module is partly fault-tolerant. It can recognize most, but not all, invalid facelet strings. If a single facelet is incorrectly identified, the **VirtualCube** will reject the string. Most pairs of incorrectly identified facelets will be similarly rejected. However, certain pairs of misidentified pairs can be accepted, although the likelihood of this happening is extremely unlikely because there are only four such pairs, and it also requires a red facelet to be identified as orange and an orange facelet to be identified as red or a green facelet to be identified as blue and a blue facelet to be identified as green. If the **VirtualCube** rejects its facelet information, the system restarts the facelet identification process.

Solving the Problems of Programming Robots Can Be Very Difficult

BRON'S
BELIEVE IT OR NOT!

Speed vs. Space

When developing software systems, it is often the case that a tradeoff exists between the amount of computation required to solve a problem and the amount of space required to implement the solution. In these cases, there are multiple options available—some faster, some smaller—that need to be evaluated before settling on a particular solution.

Such is the case with the cube solver. At one extreme, it is theoretically possible to generate the optimal solution for a Rubik's Cube in any configuration a priori and use a table lookup to find the solution to any scrambled cube. However, given that there are 12 basic moves—clockwise and counterclockwise twists of each face—and there are over 43 quintillion permutations, this solution would require an array of over 21 million 1-TB drives to store the lookup table!

At the other extreme, an algorithm can be developed that uses almost no lookup information. Such an algorithm, at the extreme, would process each cubie individually using a small repertoire of move sequences to place each cubie in the correct position and orientation. This is similar to the basic method found in the booklet that accompanies a Rubik's Cube. The drawback to this method is that it tends to generate very long sequences of moves to solve a cube because it often has to undo a previous step to make progress on a current step and then go back and fix what was undone.

Algorithms such as Thistlewaite's and Kociemba's provide a middle ground between the extremes. Each uses lookup tables that, while quite large, are able to be stored on a smart phone or other similar device. Although they do not provide optimal solutions, both algorithms generate solutions that are far shorter than the cubie-/layer-at-a-time methods.

Chapter 10

Together We Stand: The Robot Convoy

The Lost Scrolls of Robotics: #10

As soon as our respective armies take their places,
we shall correct that oversight.

—Megatron, *Transformers*

So far, we have concentrated on building teams of robots that can work together to perform intermediate-level tasks or solve problems. But what if the task is simple. What if it's just a simple matter of transportation? Do we always need a team? What if the task is simply to retrieve a container of liquid? We would like the liquid moved from point *A* to point *B*. Surely we don't need a team of communicating robots to accomplish such a simple task! We can build a nice MINDSTORMS-based mobile robot with a gripper to acquire the container from point *A* and bring it to point *B*, and voila, we're done, right? As they say, the devil is always in the details. What if we give you an approximate heading/location of the liquid but also tell you that there is more than one container at that relative location that has a liquid in it. What if we further specify that we want the container that has the green liquid in it? However, there is more than one container with a green liquid. We want the one that has a temperature between 0 to 5°C with an acidity level of 3 or less. All you have to do is take that container with the liquid we specified from point *A* to point *B*. Keep in mind that this is still basically a simple transportation problem, but consider the following challenges:

- We have an approximate heading/location for the container of liquid.
- There is more than one container at that relative location.
- We want the green liquid.
- There is more than one container with green liquid.
- We want the one with a specific temperature.
- We want the one with a specific acidity level.

What would a single MINDSTORMS-based robot look like that could deal with these challenges?

Depending on your approach, you might use the compass sensor to help find the location of the container. A color sensor, a temperature probe, an ultrasonic sensor, and a pH sensor probe all would come in handy. Well, we're in luck! HiTechnic has the color and compass sensors we need. LEGO has the temperature probe we need. The LEGO MINDSTORMS kit comes with an ultrasonic sensor. So we're good to go on the sensors. Well, we've listed five sensors so far, and the NXT brick only supports four at one time, and we left out the touch sensor. The robot has to have some way of knowing when it has grasped the container, right? So in a first run at designing a robot to accomplish our mission (our simple transportation mission), we wind up requiring at least six sensors on board. Maybe there is a more crafty way to solve this problem at the moment, but we can't think of it. So we need six sensors at least and a gripper. This is more than one NXT robot can bear.

At a minimum, therefore, we need at least two robots or one complex robot that contains more than one NXT brick. In either case, we will need teamwork and coordination between the two bricks involved. In our case, though, we have one robot that has a LEGO temperature probe, another that has a Vernier pH sensor, and another that has a robot arm/gripper and HiTechnic color sensor. Sure, we could move parts around and disassemble one of the robots to reduce the team down to two, but remember, we hate dismantling our robots! It is easier if our robots solve the problem as a team—a team of robots whose primary objective is to retrieve a container of liquid and transport that liquid from one place to another. Sometimes we call a transportation team a *convoy*. We might even call it a *robot convoy*!

Sometimes It Does Take a Team

Let's look at a project in which we create a robot convoy of three NXT robot vehicles. To be precise, a convoy can be described as "a group of vehicles organized for the purpose of control and orderly movement with or without escort protection that moves over the same route at the same time under one commander." (But in our case we won't have simple vehicles; we'll have state-of-the-art MINDSTORMS NXT–based robots!)

Typically, the purpose of a convoy is to transport vehicles and their payloads from one point to another in order to perform some specific task. While the formation of convoys is sometimes an explicitly planned operation, most often it happens in an ad hoc manner between vehicles whose drivers do not know each other, might not have common goals, and can communicate only through indirect means. Convoys can be formed and dissolved dynamically. Their life cycle can range from tens of seconds to several hours. In the formal notion of a convoy, vehicles can join and leave, and convoys themselves can split and merge. Now in our example we needed a convoy of robots because we couldn't build a single NXT-based robot that had all the functionality we needed. Usually each member of the convoy brings something special to the convoy. We could then form a convoy of these robots. The notion of a military convoy is familiar. Consider a basic military convoy. Figure 10-1 is a generic example of how a real military convoy may appear.

FIGURE 10-1 An example of a military convoy.

There is one commander in the convoy, which is always the lead vehicle. The same concept was applicable to our robot convoy. We simulated the behavior of a real convoy with three NXT-vehicle robots that traveled from one point to another. The only exception to this simulation is that no other robots joined the convoy ad hoc or left the convoy similarly. The lead robot vehicle in the convoy was designated as the team leader, and all communication was filtered through the team leader.

Using the Bluetooth Robotic-Oriented Network (BRON) for the Robot Convoy

We create a BRON piconet for our robot convoy. In this simple convoy, we have three NXT-based robots with designations C5R1, C5R2, and C5R1. The team leader (C5R1) in the convoy initiates communication with the other two team members (C5R2 and C5R3), and all communication originates from the team leader (C5R1). Each team member (C5R2 and C5R3) listens for a command to be sent from C5R1 and, on receipt of that command, executes a program. If they do not receive the command, they execute a stop program. If they receive the command, but for some reason do not execute the program, a reply is sent back to the team leader to stop the convoy. Figure 10-2 shows the relationship between the three NXT units. Table 10-1 shows a complete list of the robot vehicles and their tasks. And Figure 10-3 is a photo of our robot convoy team.

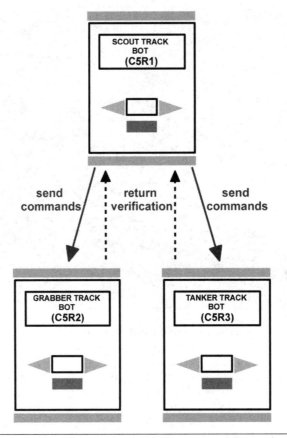

FIGURE 10-2 The relationship between C5R1, C5R2, and C5R3.

TABLE 10-1 The Robot Vehicles and Their Tasks

Robot	Tasks
Scout track bot (C5R1)	Convoy leader; sends commands to C5R2 and C5R3 to execute convoy program in regard to navigating the convoy route and what to do in case of potential danger.
Grabber track bot (C5R2)	Follows the path of C5R1; gripper hauls supplies.
Tanker track bot (C5R3)	Follows the path of C5R1; tank for convoy defense; fires projectiles on instruction from C5R1.

Each member of the robot convoy team is marked with its designation. In our robot convoy, the following robots designs were used. The first robot built (C5R1)

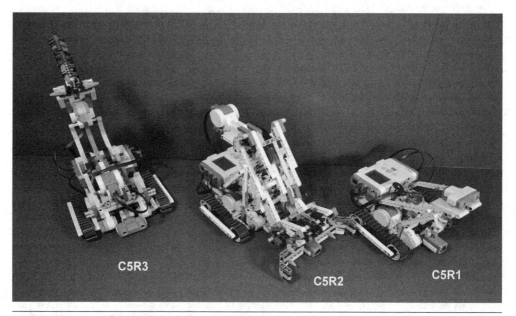

FIGURE 10-3   Photo of the robot convoy team.

is a scout track bot. C5R1 was designated as the team leader. C5R1 is equipped with an ultrasonic sensor and a color sensor in the front of the robot vehicle. As the lead vehicle in the convoy, the robot must be able to check for potential obstacles that are blocking the convoy route and be able to communicate any potential adjustments that must be made to the other robots that are following. The second robot built (C5R2) is a grabber track bot. C5R2 is the second vehicle in the convoy, following C5R1. C5R2 is equipped with a color sensor in the middle of the body of the robot vehicle and an ultrasonic sensor in the front of the robot vehicle. During the convoy, C5R2 has to be able to determine that it is still behind C5R1 after it is given and executes its program based on navigation commands from C5R1. The third robot built (C5R3) is a tanker track bot. It is equipped with an ultrasonic sensor in the front of the vehicle, which is used to detect that it is positioned behind C5R2 after it performs its navigation commands from C5R1. The convoy route is known by all robot vehicles, but the team leader instructs the other team members when it is their turn to start moving in the convoy after the team leader officially starts the convoy by moving first through the convoy route. Table 10-2 shows the capability matrix for the robot convoy.

Challenges in Robot Convoys

On first blush, moving three robots from point *A* to point *B* just seems like a simple transportation matter. However, the task of programming a convoy of autonomous

TABLE 10-2 The Capability Matrix for the Robot Convoy

Team C Members	Microcontroller	End Effectors	Mobility Type	Display	Sensors/Motors	I/O Port	Ranges	Other Capabilities
C5R1	32-bit ARM	N/A	Tractor	100- × 64-pixel liquid-crystal display (LCD)	Ultrasonic	4	0–255 cm	N/A
					Color	3	0–6	
					Servo left wheel	C	0–360 degrees	
					Servo right wheel	B	0–360 degrees	
C5R2	32-bit ARM	Gripper	Tractor	100- × 64-pixel LCD	Ultrasonic	4	0–255 cm	N/A
					Color	3	0–6	
					Touch	1		
					Servo left wheel	C	0–360 degrees	
					Servo right wheel	B	0–360 degrees	
					Servo gripper	A	0–360 degrees	
C5R3	32-bit ARM	N/A	Tractor	100- × 64-pixel LCD	Ultrasonic	4	1–255 cm	N/A
					Servo left wheel	C	0–360 degrees	
					Servo right wheel	B	0–360 degrees	

robots proved to be tricky! There were several challenges we faced. The most important challenge of all was making the robots capable of following each other without lagging behind or getting so close to each other that a collision occurred. Another challenge was programming the robots to accelerate or brake with efficiency during the convoy to avoid standstill collisions. Each robot therefore had to try to maintain a proper distance from the robot ahead. This distance can be varied according to the speed of the convoy, or it can be locked to a certain value. In the event you are duplicating this project and you have a type of convoy where each robot in the convoy is of a different build and type, similar to what we have, the maximum speed of the convoy should be limited to the top speed of the slowest vehicle in the convoy. The process of tweaking the servo motors through your programs on each vehicle in order to achieve a speed efficient enough to circumnavigate the convoy route is not an easy one. We have learned that putting the slowest vehicles in the front and the faster vehicles in the back of the convoy is a more efficient way of keeping them from colliding or running off course, as we see in Figure 10-4. Figure 10-4*a* is the correct configuration of the vehicles, slowest to fastest. Figure 10-4*b* is an incorrect configuration, fastest to slowest, which causes the convoy to fall apart.

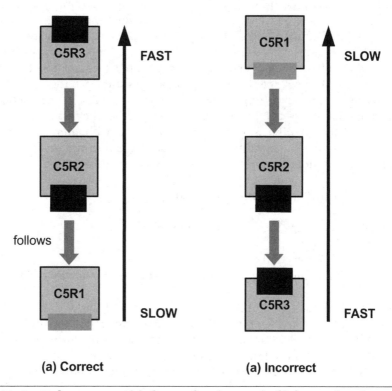

FIGURE 10-4   Robot convoy: (*a*) slow to fast (correct); (*b*) fast to slow (incorrect).

In another type of convoy, the first robot in the convoy can follow a free path. The team leader can freely navigate and let the sensors guide its path. The convoy route is not predetermined, and only the team leader will know when to stop the convoy. The robots behind must follow each robot before it. A challenge here is not to lose sight of the robot ahead. If the robot turns with a very small turning radius while moving fast, it might be impossible for the robot behind to track it, especially if there is a lag in communication. Furthermore, if the convoy consists of n robots, the first robot in the convoy must not cross its own path because it would risk driving into one of the robots further down the line.

A question that must be faced when designing a convoy is whether the robots in the convoy should follow the same trajectory. In some applications, it is critical that all the robots follow the same path as the leader because there might be some obstacles around the path. However, following the same trajectory requires that each robot knows which way the robot ahead drove. On the other hand, if the robots are allowed to "cut corners" instead of following the same trajectory, they should focus on maintaining a correct distance to the robot ahead.

The most important consideration in a robot convoy is what happens to the convoy if any robot loses connection to the convoy. Depending on the application, different safety precautions should be implemented. For example, the convoy could stop, or the robot could stop and allow the convoy to continue without it. If this is the case, then you must allow for this kind of scenario in your program by ensuring that each team member that followed the disconnected robot must be able to navigate around the disconnected robot. Also, what happens when a robot is removed from the convoy? Is the rest of the convoy capable of "filling in the gap"? As we stated previously, our convoy stopped whenever any robot lost its connection or a team member did not receive its Bluetooth message from the team leader. Recall from Chapter 1 that we discussed options when dealing with failure in a team context. We could take either the termination approach or the resumption approach. Here we chose the termination approach. That is, if any member of the team fails, we bring the whole project to a stop.

The formation of a convoy is based on a variety of factors:

- The current speed of the robot vehicles
- The desired speed
- The limitations of the robot vehicles
- The limitations of the terrain that the convoy will be navigating

The common speed of the convoy is negotiated, and the vehicles adjust their speed so that the convoy is maintained stable, preferably with uniform distances. In our robot convoy, each robot was aligned in a single-file formation (one behind the other), and we adjusted the speed of the servos powering the wheels according to each robot's speed capabilities (we will discuss the convoy speed in the next section).

Planning for the Convoy

There are five things that you must consider when you are planning to create and deploy a robot convoy:

- The limitations of each individual robot vehicle in the convoy (capability matrix)
- The terrain or surface on which the convoy will travel (a.k.a READ set)
- The convoy speed
- The weight of the robot vehicles (READ set)
- Bluetooth limitations

Limitations of Robot Vehicles

It is very important to understand the limitations of each individual robot vehicle that will participate in the convoy. There is a plethora of LEGO MINDSTORMS NXT build instructions for mobile robot vehicles. When you decide which robot vehicles you are going to build and use for a robot convoy, be mindful that there may be some flaws in the design, and you will have to either work with that flaw or try to fix the flaw before deploying the robot into the convoy. There are two different types of wheeled-movement drive systems that are prevalent in robot vehicles. The first we will refer to as a *track-steering system*. This is where one servo motor drives each track in a two-track drive system. To turn, a different amount of power is provided to the respective servo motor that powers each track. If the right servo motor has more power, the robot turns left. Likewise, if the left servo motor has more power, the robot turns right. If you give one servo motor on one side all the power and apply no power to other servo on the other side, the robot will turn either left or right in place n degrees. This is the optimal design for beginners to use because it is both easier to build and easier to design. The other system we will refer to as a *wheel-steering system*, where there is one servo motor driving the drive wheels and another servo motor that powers the steering of either the front (front-wheel drive) or rear (rear-wheel drive) wheels. Figure 10-5 shows these two types of wheeled-movement drive systems.

Probably the most common flaws in robot vehicles that are not track vehicles are issues with steering. When we built our convoy of robots, we initially didn't realize the huge number of potential issues with the steering (mechanical or steering slop) of two of our robot vehicles. We initially built five robots, three of them track vehicles and two of them wheeled vehicles. Of the two wheeled vehicles that we built, the Jeep robot vehicle was equipped with front-wheel drive, and the race-car robot vehicle was equipped with rear-wheel drive. Each wheel-steering vehicle had a considerable amount of mechanical slop. We decided not to use those robot vehicles as examples in this book, but on our website you will see the redesign of the robot vehicles to correct the issue with the steering.

One really great resource that we used was Mocpages.com. On this site, George Staples has a great tutorial about Ackermann steering geometry. Ackermann steering geometry is a steering system invented by German carriage builder George

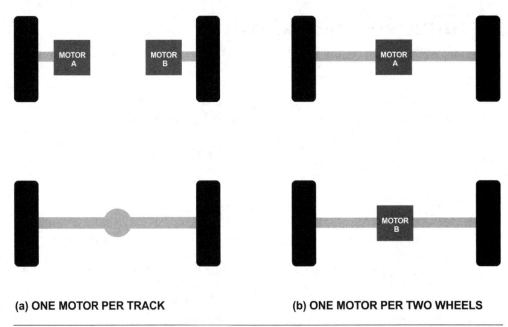

(a) ONE MOTOR PER TRACK   **(b) ONE MOTOR PER TWO WHEELS**

FIGURE 10-5 Two types of wheeled-movement drive systems: (*a*) track-steering system; (*b*) wheel-steering system.

Lankenspurger in 1817. The intention of Ackermann geometry is to avoid the need for tires to slip sideways when following the path around a curve. There is a very good Ackermann steering tutorial on Mocpages.com by George Staples specifically for LEGO MINDSTORMS NXT. You may want to investigate engineering the necessary drive system by incorporating Ackermann steering geometry into your design, as we did. Figure 10-6 shows the Ackermann steering geometry.

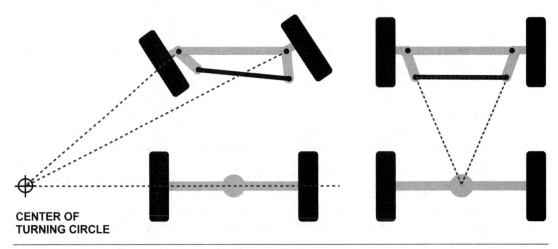

**CENTER OF
TURNING CIRCLE**

FIGURE 10-6 Ackermann steering geometry.

The Terrain (a.k.a. the READ Set)

"When military convoys are deployed in rugged terrain and challenging operational conditions, it can be difficult to maintain command and control. In mountain passes or heavily wooded areas, for example, communication can be impacted by radio propagation issues. Operational conditions also can cause disruption to the deployment of convoys due to improvised explosive devices (IEDs) and other enemy offensive tactics."[1] The terrain can be a major issue when planning for a robot vehicle convoy. One of the most difficult issues to overcome is how well the robots will navigate turns in the terrain. For example, if you chose a rugged terrain such as a gravel driveway on which to deploy your robot convoy, you would need to consider which type of robot vehicle would be able to perform best on that type of surface.

Now there's automation, and then there's automation! If our *robot environmental attribute description* (READ) set is detailed and accurate (i.e., we have a complete picture of the terrain), we can add enough programming in advance to avoid obstacles and take easier routes, etc. However, the less detailed the READ set or the less we know about the terrain, the more automation we need. We then need to add programming for edge detection, obstacle identification and avoidance, and substance identification (e.g., stay away from water). Unknown aspects of the terrain increase the complexity of convoy programming exponentially. So the more detailed and correct our READ set of the terrain is, the easier it is to write the programs for the convoy. Figure 10-7 shows the READ set for our convoy.

How Did the MINDSTORMS Convoy Perform?   We tested four surfaces or terrains for the convoy: carpet, cement, grass, and floor tile. After hours of testing, we determined the following:

- A convoy of front- and rear-wheel-drive vehicles performed poorly on carpet, grass, and floor tile but navigated well on a smooth surface.
- A convoy of tracked robots proved that they could navigate any surface or terrain.
- A convoy mixed with tracked and wheeled robot vehicles did poorly because of the deficiency of navigation performance on the aforementioned surfaces.

"From a mobility perspective, tracked vehicles offer the best solution for a versatile platform that is required to operate over diverse terrain, including extremely difficult ground, because tracks inherently provide a greater surface area than wheels."[2] For this reason, we chose to use track vehicles in our robot convoy.

Convoy Speed Concerns: Adjusting to the Speed of the Leader

In a convoy, the issue of speed is highly relative to the success of the convoy completing the convoy route. In most cases, the front vehicle of the convoy is considered the leader and should be the slowest. With that being said, you will need to adjust the speed of all the other robot vehicles in the convoy to compensate for the slowest (leader) robot in the convoy. The advantage of using this approach is that the convoy is formed more quickly and without the need of additional negotiation. Also, there is no need of a distance measurement, except the implicit one provided

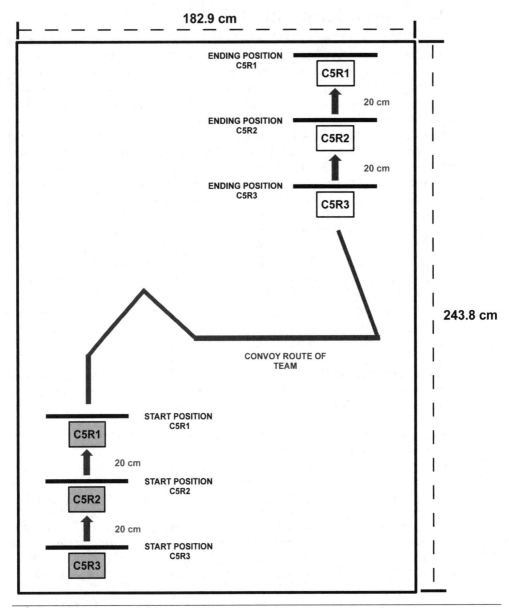

FIGURE 10-7 The READ set for the convoy of robots.

by the fact that the vehicles are in (Bluetooth) wireless range of each other. The disadvantages are that the speed of the convoy is dictated by its slowest vehicle. All the vehicles in our convoy were the same, so this was not applicable to our project. If you have a variety of robot vehicles (tracked and wheeled), you will have to spend a great deal of time configuring the servo motors on each robot vehicle that follows the next-slowest vehicle ahead of it in the convoy in order to avoid collisions and to prevent any vehicle from going off the course of the convoy route. In other words, placement of the robot vehicles in the convoy will be dictated by the slowest robot vehicle. The first robot vehicle in line will be the team leader, which will be the slowest vehicle in the convoy, followed by the next-slowest vehicle in the convoy, which will be followed by the next-slowest vehicle, and so on. This will enable the convoy to move at precise speeds and spacing, creating a more efficient convoy and reducing the risk of collisions.

All the vehicles in our convoy were of *like robots*, meaning that they each had the same build type. In this case, the average desired speed is the central point of how the convoy will navigate the course route. It would be essential for the robot vehicles in the convoy to adjust their speed to the average desired speed of the convoy before they depart the start point. In general, this strategy was perfect for our convoy. Since our convoy of three robots was of the same build, configuring the servo motors to operate at the same speed for each vehicle was simply just configuring each servo motor the same on each vehicle. The advantage of this is that we were able to determine the average desired speed and form a stable convoy that did not take long to administer. If you are using LEGO MINDSTORMS NXT-G to program your robot vehicles, you will simply just need to use the default settings in the Move block as we did. The Move block controls more than one motor at a time, and it will default to Motor B and Motor C, considering your robot design. The default power setting that controls the servo motors is set to 75 and can be seen in Figure 10-8. If you are using any other programming language to program the NXT, then you would program the servo motors accordingly.

(a) DEFAULT SETTINGS FOR SERVO MOTOR POWER

(b) NEW SETTINGS FOR SERVO MOTOR POWER

FIGURE 10-8   NXT-G's Move block configuration panel for the default settings of the servo motor power: (*a*) default and (*b*) new setting of the servo motor power.

Battery power is also a concern when determining the average speed of the convoy. The amount of battery power left to power your robot vehicles is a highly underrated factor to consider when deploying the robot vehicles in the convoy. For example, if the battery power is 5.5 in the lead robot vehicle, 7.7 in the robot following, and 9.2 in the final vehicle, even if the servo motors are configured for each robot to run at the same power, because the battery power is lower in the first and second robots in the convoy, the chance of there being a collision is highly probable. We made sure to switch out the batteries on each robot vehicle at the same time to keep the battery power in unison across the convoy.

The Weight of the Robot Vehicles

One other determining factor that can affect the convoy is the weight of the robot vehicles. Depending on which duration you choose in the Move block, the weight of the vehicle can directly affect its speed in the convoy by lowering it. There are four options for you to choose from when configuring the duration of the Move block, whether the robot is moving forward or making turns:

- Unlimited
- Degrees
- Rotations
- Seconds

Figure 10-9 shows the Move block duration configurations for (*a*) unlimited, (*b*) degrees, (*c*) rotations, and (*d*) seconds.

For our robot vehicles, we first narrowed it down to rotations and seconds and decided that the best duration option to go with was rotations because the duration it would take a heavier robot vehicle to move in time (seconds) would be slower than that of a lighter vehicle, especially if they were configured to travel at the same time. In other words, if the lead robot weighed in at 2 pounds, the next robot weighed in at ½ pound, and the third robot at 3 pounds, and they were configured to travel the same distance for the same duration in seconds, it would be highly likely that the second robot would collide with the first, even if the servo motors for each robot were powered the same.

Through hours of testing, we discovered that configuring each robot to travel a duration based on rotations lowers the possibility of a collision because the servos are going to spin the wheels or tracks n number of rotations. If each robot is configured with the same number of rotations, they will each travel that same distance.

Understanding Bluetooth Limitations

You really need to understand the Bluetooth performance limitations so that you don't end up with requirements that exceed NXT capabilities. The NXT can have three separate Bluetooth streams open on one team leader. Other team members can have only a single Bluetooth connection. However, the NXT's Bluetooth chip listens/

(a) UNLIMITED

(b) DEGREES

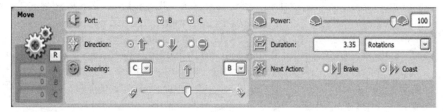

(c) ROTATIONS

(d) SECONDS

FIGURE 10-9   The Move block duration configuration panels for (*a*) unlimited, (*b*) degrees, (*c*) rotations, and (*d*) seconds.

transmits only to one stream at a time. "If one of the other two possible streams transmits to the Bluetooth chip, then the data are simply dropped. It takes 100+ ms for the NXT's Bluetooth chip to switch from one stream to another. It takes 70+ ms for a master NXT to send a Bluetooth message to a slave NXT and get a reply with the standard firmware."[3] "Although Bluetooth can provide very rapid data transfers, the particular implementation used on the LEGO MINDSTORMS NXT can cause unpredictable behavior when it's used in a convoy of NXTs."[4] "The Bluetooth protocol is implemented on the NXT as a five-message queue, which simply overflows once it's full, which will happen if incoming packets arrive faster than they can be processed."[5] When sending a Bluetooth message from the team leader to team members, remember that "each NXT has 10 mailbox numbers where wireless messages can be deposited. Each mailbox number can hold up to five messages. If the mailbox number fills up with five messages, adding one more message will cause the NXT to erase the oldest message."[6] So you will want to use multiple mailbox numbers depending on the number of messages that you will be sending from the team leader to team members.

The Robot Convoy NXT-G Program

The mobile robot vehicles were constructed with C5R1 designated as the team leader, along with the other team members, C5R2 and C5R3. Figure 10-10 is a flowchart of the convoy route process. The team leader has two sensors: one color sensor that will force C5R3 to fire to the rear if the color red is sensed and one ultrasonic sensor for sensing any obstacles at the end of the convoy route (e.g., any object will do). If the ultrasonic sensor picks up an object, it will send a message to C5R3 to start firing on it. The route itself should already be constructed. In other words, ensure that you create an environment or course where objects are already in place. The convoy must navigate the course until it reaches the finish line. We created a course, with "START" and "FINISH" positions, as well as the desired path for the convoy. Figure 10-11 is a synopsis of the movement algorithm for the convoy, and Figure 10-12 shows the NXT-G program for our robot convoy to navigate our course.

Improvement of the Robot Convoy

In the convoy project, the route was created and the robots were preprogrammed to navigate the course in order to demonstrate teamwork using communication through Bluetooth. In order to improve the project, in the future we would like to equip the robots similarly to those in the Defense Advanced Research Projects Agency (DARPA) challenge. We would like to be able to put the robots in any environment, where the convoy route is not particularly known, and the robots would have to navigate from one point to another using both Bluetooth and sensor communication. We envision having the robot leader equipped with rotating vision sensors to detect obstacles in its path and relay that information to the robots that follow. Through Bluetooth, the team leader would let the other robots in the convoy know which way to turn, based on the way the obstacle is positioned in its path. The robots that are following would make the required turn and use their sensors to determine whether they are still behind the robot in front of them when the move is made. We also would like to tackle the challenge of correcting the steering issues we encountered with the front- and rear-wheel-drive vehicles to have a more diverse convoy of robot types.

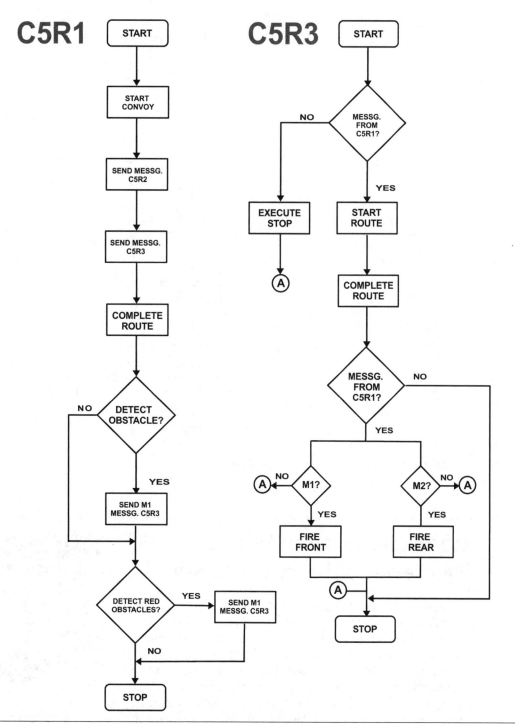

FIGURE 10-10   Flowchart of convoy route process.

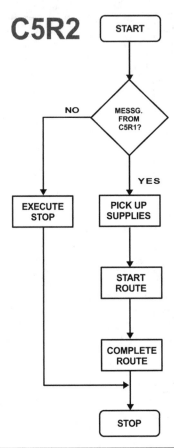

FIGURE 10-11 The movement algorithm.

Robot Cars That Can Drive by Themselves Will Never Happen

We have been intrigued by many of the Defense Advanced Research Projects Agency (DARPA) autonomous robot challenges that have taken place over the last eight years. Two of the most memorable challenges were the DARPA Grand Challenge in 2004 and the DARPA Urban Challenge in 2007. "The Grand Challenge was launched to spur innovation in unmanned ground vehicle navigation. The goal of the Challenge was to develop an autonomous robot capable of traversing unrehearsed off-road terrain."[7] "The DARPA Grand Challenge was established as a result of a

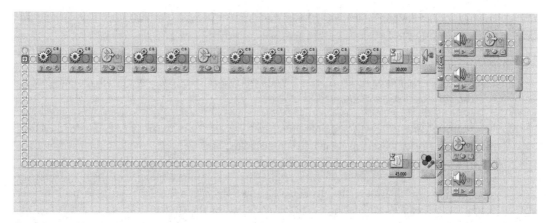

(a) C5R1 PROGRAM.

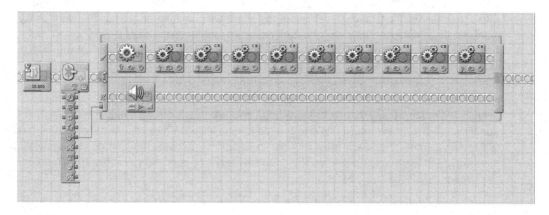

(b) C5R2 PROGRAM.

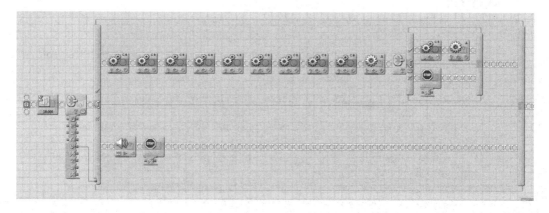

(c) C5R3 PROGRAM.

FIGURE 10-12   NXT-G program to navigate the course.

Congressional mandate, which was part of the National Defense Authorization Act, introduced in 2001, stating that:

> It shall be a goal of the Armed Forces to achieve the fielding of unmanned, remotely controlled technology such that . . . by 2015, one-third of the operational ground combat vehicles are unmanned."[8]

The first competition in 2004 required robot vehicles to navigate a 142-mile-long course through the Mojave Desert in no more than 10 hours. This was an individual competition to see which robot would be able to complete the challenge successfully. The sole goal was to see which robot vehicle would reach the finish line in a minimal amount of time. "None of the robot vehicles finished the route. Carnegie Mellon University's Red Team robot vehicle traveled the farthest distance, completing 11.78 km (7.32 mi) of the course."[9] After the challenge, we wondered whether the robots would have been more successful had they been working in teams instead of individually. We later learned that in addition to the difficulty many vehicles had with the terrain of the Mojave Desert, more important was the inability to handle two distinct problems simultaneously: sensing upcoming obstacles and following the GPS waypoints. According to DARPA Grand Challenge deputy program manager, Tom Strat, "some of the vehicles were able to follow the GPS waypoints very accurately but were not able to sense obstacles ahead. . . . Other vehicles were very good at sensing obstacles but had difficulty following waypoints or were scared of their own shadow, hallucinating obstacles when they weren't there."[10] Although we understood this was not a team competition, we couldn't help but wonder what would have transpired had the robot vehicles been in a convoy or used some form of communication among each other in order to navigate the route. If the robot vehicles would have worked in teams, would it have helped to take advantage of the strengths of each robot to compensate for the weaknesses?

Notes

1. http://rf.harris.com/media/Networked_Convoy_Rev_A_FINAL_tcm26-18152.pdf.
2. http://www.fas.org/man/dod-101/sys/land/docs/2wheels98.pdf.
3. Sivian Toledo. *Analysis of the NXT Bluetooth Communication Protocol*. http://www.tau.ac.il/~stoledo/lego/btperformance.html.
4. Ibid.
5. LEGO MINDSTORMS NXT Communication Protocol by the LEGO Group.
6. Ibid.
7. Sebastian Thrun et al. *Stanley: The Robot That Won the DARPA Grand Challenge*. http://www-robotics.usc.edu/~maja/teaching/cs584/papers/thrun-stanley05.pdf.
8. Congressional Mandate. http://archive.darpa.mil/grandchallenge04/sponsor_toolkit/congress_lang.pdf.
9. Sebastian Thrun et al. *Stanley: The Robot That Won the DARPA Grand Challenge*. http://www-robotics.usc.edu/~maja/teaching/cs584/papers/thrun-stanley05.pdf.
10. Marsha Walton. *Robots Fail to Complete Grand Challenge*. http://www.cnn.com/2004/TECH/ptech/03/14/darpa.race/index.html.

Chapter 11

The CSI Project

The Lost Scrolls of Robotics: #11
Quiet, please. I'm analyzing.

—Robby, *Forbidden Planet*

One of the major focus points of this book is on executing tasks by using a team of robots. But there is another aspect of robots and teams of robots that we haven't spent much time on yet—specifically, the idea that robots can do what humans can't do or shouldn't do. Robots can go into situations that are too dangerous for humans or into environments that are not suitable for human life.* Robots can handle materials that are too heavy or too dangerous for human beings. It's in these situations that teams of robots prove their value beyond any doubt.

In this book we built several teams of robots. Each team was designed to execute or solve some scaled-down version of a real-world task or problem. Each team was working in a modeled world. Our emphasis was on demonstrating the techniques of communication and coordination necessary for a team of robots to accomplish its task rather than giving you full-blown robot applications where the complexity of the project would hide the fundamentals of the communication and cooperation techniques we have presented. These techniques are the basis of any team. For the crime scene investigation (CSI) project, we want to highlight that aspect of robots where the team of robots is handling materials and environments that are potentially dangerous for humans. Once again, to focus on the fundamentals, we use a modeled world. But it should be easy to imagine the real scenario for your team of robots.

Overview of the CSI Project

In this project, it's not only the teamwork on which we want to focus but also the automation that is involved. This project does not use any remote controls to control

---

*We were in complete awe at the recent landing of the robot Curiosity on the surface of Mars. If you look at Curiosity, it's not very different from some of the simple robots we've built in this book. It has six wheels and a host of sensors, much like our own D1R1 and A3R3 Tetrix-based builds.

any of the robots. All robot actions are taken as a result of programming and onboard robot decision making. The robots are operating within the context of a scenario. This is the scenario: a top-secret secure warehouse, deep in the middle of nowhere, and a sentry robot that is designed to patrol and monitor activity and materials. The sentry robot is part of a CSI team whose job it is to investigate, secure, and report any anomalies that the sentry robot encounters. The contents and location of the warehouse are dangerous and are considered classified. The sentry robot makes its rounds of the warehouse randomly. It uses *radio frequency identification* (RFID) to identify every object in the warehouse. During one of its rounds, the sentry robot encounters an unknown container, performs an immediate analysis, and sends for more members of the CSI team. The CSI members travel to the warehouse in a convoy. You guessed it, the convoy from our convoy project. Once onsite, one member of the CSI team uses its robotic arm to secure the liquid. The CSI members perform a more in-depth analysis and determine that the container contains an unknown liquid. If the substance is harmless, a report is made using an Android smart phone. If the substance is identified as harmful, an emergency call is placed to several agencies of concern. Now, of course, this is a mock situation, but it should not be hard to believe that our mock situation does have real-world counterparts.

The entire investigation is more or less triggered by our RFID sensor system that the sentry robot uses. The RFID sensor that we used for this project is made by CODATEX and is supported by NXT-G and leJOS. This system and how it works were discussed in Chapter 3, which discussed sensors. In our mock scenario, a basic sentry robot equipped with an RFID sensor saves the day.

The Tasks and Problems Encountered in Warehouse X

We've given a sketch of the scenario and one small part of the processing, but there are many serious robotic challenges that face the CSI project in Warehouse X:

- How does the robot sentry patrol the warehouse?
- How does it cover the floor space of the warehouse? Line following? Edge detection? Exhaustive search?
- If it is using RFID tags to identify/verify the contents of the warehouse, this may work fine when determining whether something is missing or for counting items, but in this case there is something in the warehouse that doesn't belong, and it has no tag. How does the sentry identify a new object?
- What are the object detection options?
- Once the unidentified object is found, how can the robot sentry describe the object's location, size, weight, etc. to the primary CSI team?
- Because of the status of Warehouse X, there is little or no lighting. What robot capabilities will the CSI team need to identify the unknown object?
- Does the robot sentry request help of the entire team or part of the team?
- If the sentry requests only part of the team, which part will it request and why?

- Once the robot sentry does get the attention of the CSI team, what will it communicate about the unknown object?
- Not all robots have the same sensors. In fact, no one else on the team has RFID capabilities outside of the sentry. How do other team members identify the unknown object?
- In our scenario, we've let on that the unknown object is some kind of container that contains an unknown liquid. How was this fact determined?
- What robot capabilities are needed to determine if something is a container or if something is a liquid or solid?
- Are the challenges to the team of robots clear?

The Capability Matrix of the CSI Project

Throughout this book we introduced the capabilities of the robots that ultimately make up the CSI team. We've built LEGO MINDSTORMS basic scout bots, Tetrix-based range rovers, and an Android-powered Rubik's Cube solver. We've used the compass, ultrasonic sensors, and temperature sensors. We deployed Tetrix-based direct-current (DC) motors and servos for our robot actuators. We included a project that demonstrated the notion of a convoy of robots that move uniformly from one location to another. In fact, we showcased bits and pieces of the CSI team throughout the book. The communication and coordination techniques that were demonstrated can be used to address some of the challenges presented in Warehouse X. Table 11-1 shows the capability matrix of the robots we have.

While this capability matrix shows all the robotic/nonrobotic capability that we have at our disposal, it does not necessarily show which robots/devices will make up the CSI team. Keep in mind that a single BRON or piconet, depending on how it is configured, can have from four to at most seven members. We do not consider scatternets or interfaced BRONs in this book, so we are limited to a team of four to seven members. Which robots do we use? How do we map the sensors and capabilities that we have in the capability matrix to the problems/challenges of Warehouse X and to the general description of the problem scenario? Recall the READ set. Before we can really decide which robots make the team, we really need to describe in as much detail as we can the READ set of Warehouse X.

The READ Set of Warehouse X

In this mock scenario, Warehouse X has a floor space of 6 × 8 ft. It will be divided into four 2- × 6-ft quadrants named quadrant 1, quadrant 2, and so forth. There will be at least six objects dispersed through the quadrants. Some of the objects will have RFID transponders. One of the objects will be a 5- × 7-cm container, and another object will be a 6- × 9-cm container. Warehouse X can only be entered by one of its two 6-ft edges. The warehouse has a height of 7 ft. The color of the warehouse floor is black. The warehouse's floor is semirugged. The warehouse is dimly lit. The CSI primary team is located 4 ft away from Warehouse X. There is a straight, unimpeded trajectory from the CSI primary team to Warehouse X. The contents of the containers

TABLE 11-1 Capability Matrix of All the Robots Used in This Book

Team Members	Microcontroller	End Effectors	Mobility Type	Display	Sensors/Motors	I/O Port	Range	Other
Team B Members								
Twisty (B1R2)	32-bit ARM	Upper: Twister (rotating platform oriented upside-down)	N/A	100- × 64-pixel liquid-crystal display (LCD) screen	Upper twister	A	0–360 degrees	N/A
		Lower: Twister (rotating platform)			Lower twister	B	0–360 degrees	
		Aligner: Two fingers moving in parallel			Aligner	C	−2,300–0 degrees	
Flippy (B1R1)	32-bit ARM	Flipper	N/A	100- × 64-pixel LCD screen	Flipper (linear actuator)	A	−120–0 degrees	N/A
		Gripper			Gripper (linear actuator)	B	−45–0 degrees	
Android	32-bit dual-core ARM		N/A	800 × 400 WVGA	Camera	N/A	N/A	N/A
Team C Members								
C5R1	32-bit ARM	N/A	Tractor	100- × 64-pixel LCD	Ultrasonic	4	0–255 cm	N/A
					Color	3	0–6	
					Servo left wheel	C	0–360 degrees	
					Servo right wheel	B	0–360 degrees	
C5R2	32-bit ARM	Gripper	Tractor	100- × 64-pixel LCD	Ultrasonic	4	0–255 cm	N/A
					Color	3	0–6	
					Touch	1	0–1	
					Servo left wheel	C	0–360 degrees	
					Servo right wheel	B	0–360 degrees	
					Servo gripper	A	0–360 degrees	

Team Members	Microcontroller	End Effectors	Mobility Type	Display	Sensors/Motors	I/O Port	Range	Other
C5R3	32-bit ARM	N/A	Tractor	100- × 64-pixel LCD	Ultrasonic	4	0–255 cm	N/A
					Servo left wheel	C	0–360 degrees	
					Servo right wheel	B	0–360 degrees	
Team D and A Members								
D3C3	AMD Turion 64-bit dual 2.8-GHz processor	N/A	Portable	19-in HD LCD	N/A	N/A	1.5 TB of storage	802/11b wireless Bluetooth USB 10/100 Ethernet Modem, infrared
D1R1	32-bit ARM7, 64-kB RAM	Arm and gripper	Four wheels	100- × 640-pixel LCD screen	LEGO touch	3	1–0	Bluetooth, USB
					HiTechnic color	2	16 colors	
					LEGO ultrasonic	4	0–255 cm	
					DC motor controller	1	360 degrees	
					Left wheel DC motor	DC controller 1-, 1+	360 degrees	
					Right wheel DC motor	DC controller 2-, 2+	360 degrees	
					Left wheel encoder	DC motor controller enc. 1	360 degrees	
					Right wheel encoder	DC motor controller enc. 2	360 degrees	
					Arm servo	Servo controller channel 1	90 degrees	
					Gripper servo	Servo controller channel 2	90 degrees	

(continued on next page)

TABLE 11-1 Capability Matrix of All the Robots Used in This Book (*continued*)

Team Members	Microcontroller	End Effectors	Mobility Type	Display	Sensors/Motors	I/O Port	Range	Other
D1R2	32-bit ARM7, 64-kB RAM	Probe and touch sensor extension	Two tractor wheels	100- × 64-pixel LCD screen	LEGO touch	1	1–0	Bluetooth, USB
					Vernier pH sensor adapter	3	pH 1–14	
					Probe servo	B	90 degrees	
					Right tractor wheel servo	C	360 degrees	
					Left tractor wheel servo	A	360 degrees	
A3R3	32-bit ARM7, 64-kB RAM	Arm with probe effector	Three wheels	100- × 64-pixel LCD screen	LEGO temperature probe	2	−20–120°C, −4–248°F	Bluetooth, USB
					Ultrasonic	4	0–255 cm	
					HiTechnic compass	3	0	
					DC motor right wheel	DC motor controller motor 1−, 1+	360 degrees	
					DC motor left wheel	DC motor controller motor 2−, 2+	360 degrees	
					Right wheel encoder	DC motor controller enc. 1	N/A	
					Left wheel encoder	DC motor controller enc. 2	N/A	
					DC controller	1	N/A	
					Right tractor wheel servo	C	360 degrees	
					Left tractor wheel servo	A	360 degrees	
					Servo motor for arm	Servo Controller channel 2	180 degrees	

Team Members	Microcontroller	End Effectors	Mobility Type	Display	Sensors/Motors	I/O Port	Range	Other
A1R1	32-bit ARM7, 64-kB RAM	N/A	Tractor wheel	100- × 64-pixel LCD screen	Ultrasonic	4	0–255 cm	Bluetooth, USB
					CODATEX RFID	1	5–12 bytes ID	
					LEGO color	3	0–6 colors	
					Servo right wheel	B	0–360 degrees	
					Servo left wheel	C	0–360 degrees	
					Servo center	B	0–360 degrees	
Android smart phone	32-bit dual-core ARM (Android)	N/A	N/A	800 × 400 WVGA	Camera	N/A	N/A	Bluetooth, WiFi

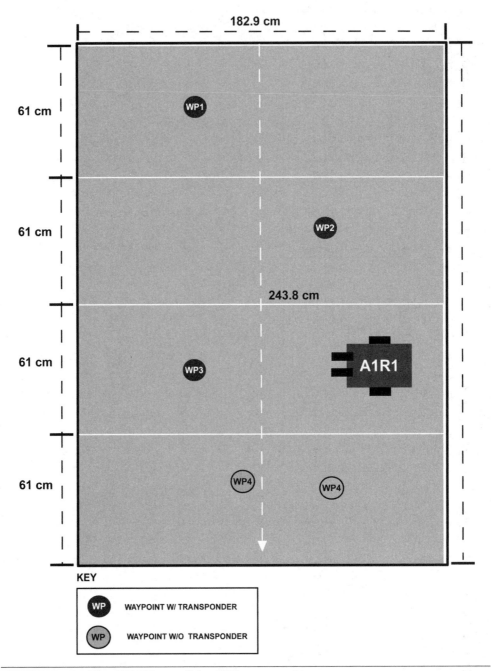

FIGURE 11-1 The READ set of the CSI team project.

are determined by the RFID tag. Some of the objects in Warehouse X are containers; some are not. Any container in Warehouse X will contain some kind of matter (e.g., solid, liquid, gas, or plasma*).

We look at the problem/scenario that we need the robots to solve or operate within by looking at the READ set in Figure 11-1. We take an assessment of our robot capabilities using the capability matrix, and then we put together our best team of robots to get the job done. Figure 11-2 is a photo of the CSI team that we will use to deal with the problem of Warehouse X.

Notice that the team is heterogeneous; it consists of a mix of Bluetooth-enabled devices. We have basic Tetrix-based robot designs, as well LEGO MINDSTORMS robot designs. In addition to the four robots, the team also consists of a notebook computer and an Android smart phone. The number of devices involved should suggest how our piconet is to be organized. The notebook computer and the Android smart phone give the CSI team the added computational and communication power that is needed. The names of the robots and computers in this project are

- D1R1
- A3R3
- D1R2

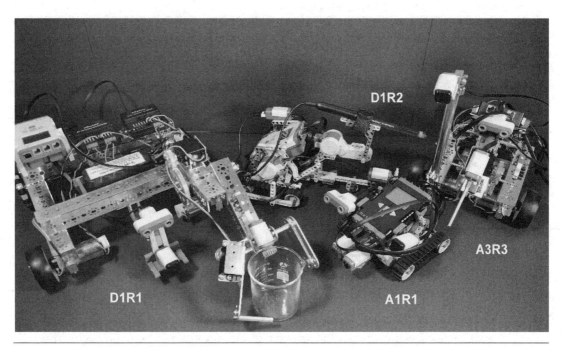

FIGURE 11-2 Photo of the CSI team.

---

*Imagine what the Robot Curiosity has to deal with on Mars.

- A1R1
- D3C1
- D3C2

Table 11-2 lists the basic functions of the robots within this scenario. Recall the naming convention that we've used throughout this book. Now once the team is selected, we have a basic plan to deal with the scenario in Warehouse X. We represent our plan graphically in a flowchart so that we can criticize and evaluate it before we actually program the robots and burn up power sources and motors on trial and error.

TABLE 11-2 Basic Functions of the Robots and Devices on the CSI Team

Robot/Device	Tasks
D3C1	Computer as the team leader; communicates with all team members; coordinates action-based contents of Bluetooth message communications.
D3C2	Smart phone; receives Bluetooth message and makes emergency calls.
D1R1	Transports containers.
D1R2	Performs pH analysis.
A1R1	Scout bot; uses RFID to check containers at location; signals when there is an unknown container present.
A3R3	Performs checks on the temperature of the substance in the container.

An Approach to Solving the CSI Warehouse X

The flowchart (Figure 11-3) is an excellent means of organizing our thoughts prior to the task of programming our robots and other devices. First, the visual picture in most cases will be much easier to understand than the code. It is usually easier to spot errors and possible design flaws in our logic if we use a visual design tools such as flowcharts. It's easier to have others look at and evaluate our designs and possibly discuss potential alternatives when everything is in a visual graphical format. Figure 11-4 is the class relationship diagram for the CSI project. This shows the relationships (containment and inheritance) between all the robot team members and their support classes.

There are four primary robots used for the CSI project. Each robot has an associated Java class. The Java class is for all intents and purposes the "software version" of the robot. For every hardware component of the robot, there is a software component. We'll take a look at each of the Java class declarations for the robots and the initialization routine that starts each robot up. Listing 11-1 contains the major components of `a1r1`, the robot sentry that patrols Warehouse X.

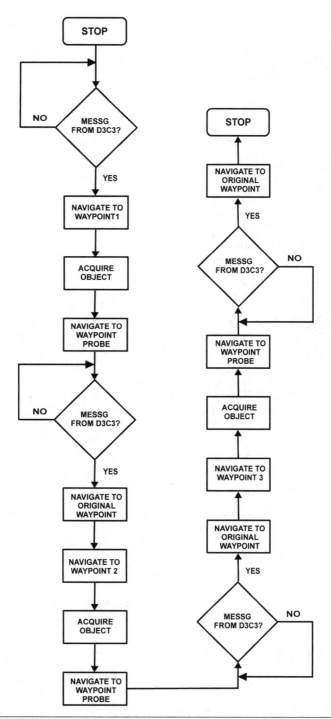

FIGURE 11-3   Flowcharts for CSI team project: (*a*) D1R1.

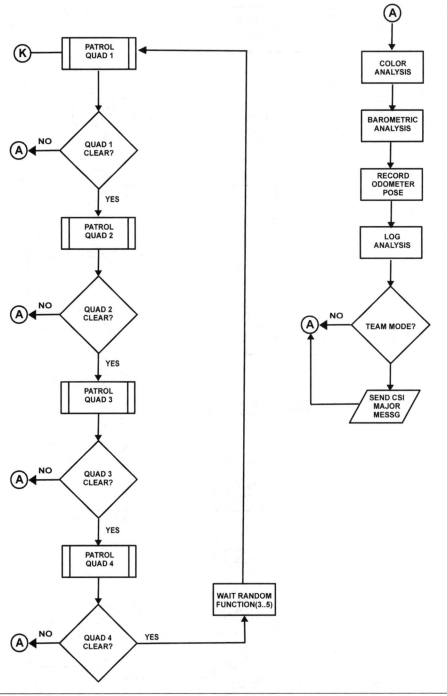

FIGURE 11-3 Flowcharts for CSI team project: (*b*) complete robot sentry processing.

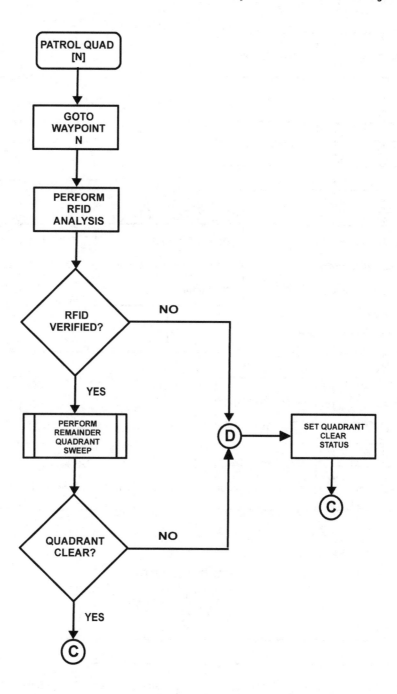

FIGURE 11-3   Flowcharts for CSI team project: (*c*) complete robot sentry processing continued.

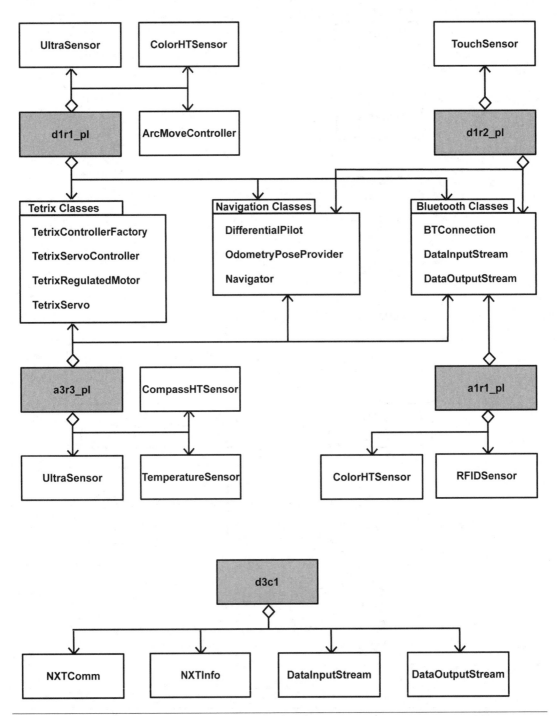

FIGURE 11-4 The class relationship diagram for the CSI project.

LISTING 11-1

```
 1  import java.io.DataInputStream;
 2  import java.io.DataOutputStream;
 3  import lejos.nxt.UltrasonicSensor;
 4  import lejos.robotics.Color;
 5  import lejos.nxt.ColorSensor;
 6  import lejos.nxt.addon.RFIDSensor;
 7  import lejos.nxt.*;
 8  import lejos.nxt.SensorPort;
 9  import lejos.nxt.LCD.*;
10  import lejos.nxt.comm.BTConnection;
11  import lejos.nxt.comm.Bluetooth;
12  import lejos.robotics.navigation.DifferentialPilot;
13  import lejos.robotics.localization.OdometryPoseProvider;
14  import lejos.robotics.navigation.Pose;
15  import lejos.robotics.navigation.Navigator;
16  import java.lang.Byte;
17
18  public class a1r1_p1
19  {
20      public UltrasonicSensor Vision;
21      public ColorSensor RobotColorSensor;
22      public RFIDSensor RobotRfidSensor;
23
24      int ObjectColor;
25      double WheelDiameter;
26      double TrackWidth;
27      DifferentialPilot a1r1Pilot;
28
29      OdometryPoseProvider Odometer;
30      Navigator a1r1Navigator;
31      Pose CurrPos;
32      Pose ObjectPos;
33      int OneSecond = 1000;
34      int NoObjectVisible = 255;
35      Sound AudibleStatus;
36
37      BTConnection Btc;
38      DataInputStream dis;
39      DataOutputStream Dout;
40
41
42
43
```

```
44      public alr1_p1() throws InterruptedException
45      {
46          Vision = new UltrasonicSensor(SensorPort.S4);
47          RobotColorSensor = new ColorSensor(SensorPort.S3);
48          RobotRfidSensor = new RFIDSensor(SensorPort.S1);
49          LCD.drawString("Sensors constructed",0,0);
50          WheelDiameter = 3.0;
51          TrackWidth = 13.0;
52          LCD.drawString("motors Constructed",0,0);
53          alr1Pilot = new
                        DifferentialPilot(WheelDiameter,TrackWidth,
                        Motor.C,Motor.B);
54          alr1Pilot.setTravelSpeed(6);
55          alr1Pilot.setRotateSpeed(30);
56          LCD.drawString("Pilot Constructed",0,0);
57          Thread.sleep(3000);
58          CurrPos = new Pose();
59          CurrPos.setLocation(0,0);
60          Odometer = new OdometryPoseProvider(alr1Pilot);
61          Odometer.setPose(CurrPos);
62          alr1Navigator = new Navigator(alr1Pilot,Odometer);
63          alr1Navigator.singleStep(true);
64          LCD.drawString("Odometer Constructed",0,0);
65          Thread.sleep(5000);
66          RobotRfidSensor.wakeUp();
67
68
69      }
70      //...
71
72  }
```

Lines 3 through 6 declare the basic sensors that the robot has. Lines 8 through 14 show the navigation, pilot, and odometer classes that the robot sentry uses. These classes allow the sentry to travel to waypoints easily to track its current location, as well as navigating to its normal patrol points. Lines 44 through 66 are the initialization routine that performs the basic setup for the sentry robot. Recall that the sentry robot travels to five designated waypoints, and at each waypoint, it uses the RFID sensor to determine whether the contents of the warehouse are where they are supposed to be. Listing 11-2 shows how we accessed the RFID sensor.

LISTING 11-2

```
1
2   public void readTransponder(String TP) throws Exception
```

```
 3  {
 4
 5      RobotRfidSensor.readTransponderAsLong(false);
 6      Long SValue = new
                    Long(RobotRfidSensor.readTransponderAsLong(false));
 7      Thread.sleep(5000);
 8      TP = new String(SValue.toString());
 9
10
11
12  }
13
```

There are three other Java methods for `a1r1` that are critical to its patrolling operation. In addition to use of the RFID, the quadrant scanning and object preliminary analysis are important. If the RFID sensor does not see an appropriate transponder, it will return a 0. Each of the recognized items in Warehouse X has a transponder attached to it. Each transponder has a separate ID. The transponder ID is typically a long number. For example, one of the readings from the transponder that we used was

1045396127824

The three primary operations of the sentry robot are `quadrantScan()`, `analyzeQuadrant()`, and `preliminaryObjectAnalysis()`. These three methods, shown in Listing 11-3, represent the primary work of the sentry robot.

LISTING 11-3

```
 1
 2
 3
 4  ...
 5
 6  public void quadrantScan(int Width, int Length,int QuadSize) throws
    Exception
 7  {
 8
 9      int Search = 0;
        int Angle = -30;
10      boolean SectorOk = true;
11      int Circumference = (Math.abs(180 / Angle) - 1);
12      while(SectorOk && (Search < QuadSize))
13      {
```

```
14
15          // One circumferemce
16          for(int X = 0; X < Circumference; X++)
17          {
18              rotate(Angle);
19              waitForRotate(Angle);
20              SectorOk = analyzeQuadrant(Length);
21              if(SectorOk){
22                  rotate(Angle);
23                  waitForRotate(Angle);
24                  SectorOk = analyzeQuadrant(Width);
25              }
26          }
27          Search++;
28          Width = Width + 5;
29          Length = Length + 5;
30
31      }
32
33  }
34
35
36
37
38  public boolean analyzeQuadrant(int Length) throws Exception
39  {
40
41      int ObjectLocation = -1;
42      int DistanceToTravel = 0;
43      travel(Length);
44      waitUntilStop(Length);
45      ObjectLocation = getUltrasonicDistance();
46      Thread.sleep(OneSecond * 2);
47      if(ObjectLocation == NoObjectVisible){
48          return true;
49      }
50      else{
51
52          AudibleStatus.beep();
53          Thread.sleep(OneSecond);
54          if((ObjectLocation != NoObjectVisible) &&
                ((ObjectLocation >= 1) &&
                (ObjectLocation <= 12)))
                {
55              DistanceToTravel = ObjectLocation - 6;
```

```
56                    travel(DistanceToTravel);
57                    waitUntilStop(Length);
58                    soundAlert();
59                    preliminaryObjectAnalysis();
60                    LCD.drawInt(ObjectLocation,1,1);
61                    Thread.sleep(OneSecond * 10);
62                    return(false);
63
64                }
65
66
67        }
68    return true;
69
70
71 }
72
73
74 void preliminaryObjectAnalysis() throws Exception
75 {
76
77        String ObjectX = new String("NotValid");
78        startFirmware();
79        Thread.sleep(OneSecond * 5);
80        readTransponder(ObjectX);
81        LCD.clear();
82        LCD.drawString(ObjectX,0,0);
83        AudibleStatus.twoBeeps();
84        Thread.sleep(OneSecond);
85        ObjectPos = Odometer.getPose();
86        AudibleStatus.beep();
87        Thread.sleep(OneSecond);
88        ObjectColor = RobotColorSensor.getColorID();
89        AudibleStatus.beep();
90        Thread.sleep(OneSecond * 3);
91        soundAlert();
92
93
94
95 }
```

The size of the quadrant that the sentry searches depends on the parameters Length, Width, and QuadSize, which are shown in line 6. The sentry's analysis performed in lines 80 through 88 represents the preliminary CSI analysis that is performed by the sentry. This analysis is sent to D3C1 using Bluetooth and will be part of the formal analysis that determines what the unknown substance is or is not. Keep in mind that D3C1 is the team leader. The team leader in this case is a computer and uses the **NXTComm** class to define the communications between the robots. Listing 11-4 shows the basic class definition of the team leader and how to set up four team members using the Bluetooth protocols.

LISTING 11-4 Team Leader

```
 1    // Listing 11.4 Team Leader
 2
 3
 4    import java.io.DataInputStream;
 5    import java.io.DataOutputStream;
 6    import java.io.BufferedReader;
 7    import java.io.InputStreamReader;
 8    import lejos.pc.comm.NXTCommLogListener;
 9    import lejos.pc.comm.NXTConnector;
10    import lejos.pc.comm.NXTInfo;
11    import lejos.pc.comm.NXTComm;
12    import lejos.pc.comm.NXTCommFactory;
13    import lejos.pc.comm.NXTCommBluecove;
14
15
16    public class d3c1{
17
18        String D1R1 = "D1R1";
19        String A3R3 = "A3R3";
20        String D1R2 = "D1R2";
21        String A1R1 = "A1R1"
22
23        float Temperature[];
24        DataOutputStream Dout;
25        DataInputStream Din;
26        int OneSecond = 1000;
27
28        NXTComm D1R1Comm;
29        NXTComm A3R3Comm;
30        NXTComm D1R2Comm;
31        NXTComm A1R1Comm;
32
33        NXTInfo D1R1Robot;
34        NXTInfo D1R2Robot;
35        NXTInfo A3R3Robot;
36        NXTInfo A1R1Robot;
37
```

```
38    public d3c1() throws Exception
39    {
40      Temperature = new float[3];
41      D1R1Comm = NXTCommFactory.createNXTComm(NXTCommFactory.
                    BLUETOOTH);
42      A3R3Comm = NXTCommFactory.createNXTComm(NXTCommFactory.
                    BLUETOOTH);
43      D1R2Comm = NXTCommFactory.createNXTComm(NXTCommFactory.
                    BLUETOOTH);
44      A1R1Comm = NXTCommFactory.createNXTComm(NXTCommFactory.
                    BLUETOOTH);
45
46      if(D1R1Comm == null){
47
48        System.out.println("Could Not find Brick");
49        System.exit(1);
50
51      }
52      else{
53
54        System.out.println("Found Brick" + D1R1);
55
56      }
57      D1R1Robot = new   NXTInfo(NXTCommFactory.BLUETOOTH,D1R1,
                    "00:16:53:0C:54:91");
58      A3R3Robot = new   NXTInfo(NXTCommFactory.BLUETOOTH,A3R3,
                    "00:16:53:01:A9:8D");
59      A1R1Robot = new   NXTInfo(NXTCommFactory.BLUETOOTH,A1R1,
                    "00:16:53:0A:7F:06");
60      D1R2Robot = new   NXTInfo(NXTCommFactory.BLUETOOTH,D1R2,
                    "00:16:53:14:1A:F3");
61      if(D1R1Robot == null || A3R3Robot == null){
62
63        System.out.println("Could not create NXTInfo");
64        System.exit(1);
65      }
66      else{
67
68          System.out.println("Created NXTInfo: " + D1R1);
69
70
71      }
72
73  }
```

Notice that lines 40 through 44 and lines 57 through 60 contain the code that sets up the Bluetooth communication with the CSI team. Recall that the utility `hciscan` tool is used to get the actual addresses of the robots. The addresses of the robots are shown in lines 57 through 60 and are represented by a string such as:

```
00:16:53:14:1A:F3
```

Also keep in mind that once the Bluetooth connection is made, the `DataInputStream` and `DataOutputStream` classes are used to communicate between robots and from robot to team leader. For example:

```
public void getMessageD1R1(int Message) throws Exception
  {

        Din = new DataInputStream(D1R1Comm.getInputStream());
        Message = Din.readInt();
        System.out.println("message from D1R1 Received " + Message);
        Thread.sleep(OneSecond * 2);
        Din.close();

  }


    public void sendMessageA3R3(int Message) throws Exception
    {

        Dout = new DataOutputStream(A3R3Comm.getOutputStream());
        Dout.writeInt(Message);
        System.out.println("Message sent to A3R3 " + Message);
        Thread.sleep(OneSecond * 2);
        Dout.close();


    }
```

The `sendMessage()` writes an integer to robot A3R3 over a Bluetooth connection, and the `getMessage()` reads an integer from its Bluetooth connection. The messages sent from A1R1 contain information about the substance that needs to be analyzed. For this CSI project, we use the Vernier probe to do the final analysis. The acidity level of the substance found in Warehouse X will determine whether a message is sent to D3C3 (the Android phone). Listing 11-5 shows how we used the Vernier pH probe in the CSI project with leJOS. It is important to note that the leJOS environment does not have the Vernier pH probe simplified class called `PhProbe`. Instead, we had to directly access `SensorPort.S2`.

LISTING 11-5

```
1   //(RawReading / 2^10) * 5 = voltage
2   // voltage * (-3.838) + 13.720 = measurement
3   // -3.838 is slope
4   // 13.720 is intercept
5
6   public double phProbeVoltage(double ActualReading)
7   {
8       double Voltage = 0;
9       double Measurement = 0;
10      Voltage = ((ActualReading / Math.pow(2,10)) * 5);
11      Measurement = Voltage * (-3.838) + 13.720;
12      return(Measurement);
13
14  }
15
16
17  void phAnalysis()
18  {
19
20      LCD.clear();
21      double RawValue = 0;
22      double Measurement = 0;
23
24
25      SensorPort.S2.setTypeAndMode(SensorPort.TYPE_LIGHT_ACTIVE,
                                    SensorPort.MODE_RAW);
26      RawValue = SensorPort.S2.readRawValue();
27      Double RV = new Double(RawValue);
28      System.out.println(Math.round(RV.doubleValue()));
29
30      Measurement = phProbeVoltage(RawValue);
31      Double PhValue = new Double(Measurement);
32      System.out.println(Math.round(PhValue.doubleValue()));
33
34
35
36
37  }
```

Notice on line 25 that we had to set **RAW** mode. The values that are read from the sensor then must be converted to the actual pH values using the **type** and **mode**. We experimented with other types; the deciding factor, as of this writing, was setting the sensor **phProbeValue()**. In this chapter we show only some of the more interesting methods used in the CSI project. The complete source code for the CSI project and all of the major projects in this book can be downloaded from the website (www.robotteams.org).

In addition to leJOS, many of the projects are available in NXT-G, NXC, and LabVIEW.

Summary of the CSI Project

The challenges encountered in Warehouse X are just some of the challenges that the team of robots will face in solving the mystery. Solving these problems are so fundamental to all kinds of robotic tasks and to robot programming in general that Ctest Laboratories and Northeast Ohio Association for Computing Machinery (NEOACM), in conjunction with Youngstown State University, are currently organizing a series of CSI/CLUE robotic competitions that will face the CSI/CLUE challenges of Warehouse X. In this book, we've presented a scaled-down version of the Warehouse X problem with some simplifying assumptions, but we leave enough intact to capture the value of a team of robots working autonomously to solve a problem or execute a task.

Robots Will Take Over the World One Day

BRON'S BELIEVE IT OR NOT!

From the golems of ancient Judaic myth, who undoubtedly were premonitions of the ominous rogue robot, to Tilotttama, the humanoid of antiquity documented eons ago in Sanskrit myth, to the mechanical duck (perhaps the first true automaton) of Jacques de Vaucanson presented to the Academie Royale des Sciences in Paris in 1738, it seems that there has always been a fear that one day we would become the victims of our own preoccupation with the invention of mechanical versions of things that usually are reserved for and exclusively within the providence of creation by Nature. The fear that an army of automated mechanical men and devices will one day reduce humanity to a life of servitude is commonly believed among cultures the world over. Nothing captures the fear better than the popular 2005 motion picture *I Robot*, in which household robots, through the cajoling of the artificial intelligence VIKI, uniformly decide to turn on their owners and plan a revolt that changes the balance of power between robots and humans in one night!

Of course, being taken over by an army of disgruntled robots is not the likely scenario. Robots are still far too dependent on us for power sources, parts, maintenance, and programming. While there are some robots that repair other robots, the whole robotic life cycle is still under our complete control. But that doesn't mean that we won't succumb to robots anyway. Rather than being taken over by robots, we are willingly, little by little, giving our automation permission. It's not the typical humanoid robot that we have to worry about;

instead, it's a thousand little pieces of automation that we freely give our lives over to. Such things as the automatic teller machine (ATM) that handles our monetary and financial transactions for us. We've grown accustomed to the metal/plastic face of the ATM and prefer it over the smiling, courteous human counterpart. The automated grocery cashier that prices and checks out our groceries on behalf of (or instead of) its human counterpart. The latest model luxury car that is capable of parking itself, dispensing with the need for a valet. Car, park thyself! The state-of-the-art laser-driven carwash that uses computer programming and automation to clean your vehicle. So much for turtle wax and buckets of soap! We gladly let unmanned drones get the military job done, automating what used to be done by human pilots. The new model dishwashers are a sight to behold. There is nothing menacing about an automated dishwasher, but think of the jobs that it replaces. Dishwashing used to be an honorable occupation, that is, before the machines took over! No, it's not true that robots will take over the world one day; on the contrary, one day we'll look around and realize that we have voluntarily, willfully, and without force, in the name of convenience, efficiency, and perhaps cost, given the whole world over to them.

Appendix A

Standard Java Classes for leJOS Bluetooth

Standard Java Classes

Class `DeviceClass`

Description

The `DeviceClass` class represents the class of device (CoD) record as defined by the Bluetooth specification. This record is defined in the Bluetooth Assigned Numbers Document and contains information on the type of device and types of services available on the device. The Bluetooth Assigned Numbers Document (www.bluetooth.org/assigned-numbers/baseband.htm) defines the service class, major device class, and minor device class.

DeviceClass

Prototype	Returns	Parameters	Exception
`DeviceClass(int record)`	None	`int record`	Throws: `IllegalArgumentException`—if record has any bits between 24 and 31 set

getServiceClasses

Description: Retrieves the major service classes. A device may have multiple major service classes. When this occurs, the major service classes are bitwise OR'ed together.

Prototype	Returns	Parameters	Exception
int getServiceClasses()	int—the major service classes 0x22000—networking and limited discoverable major service classes 0x100000—object transfer major service class	None	Throws: None

getMajorDeviceClasses

Description: Retrieves the major device class. A device may have only a single major device class.

Prototype	Returns	Parameters	Exception
int getMajorDeviceClasses()	int—the major device class 0x00—miscellaneous major device class 0x200—phone major device class	None	Throws: None

getMinorDeviceClasses

Description: Retrieves the minor device class. A device may have only a single minor device class.

Prototype	Returns	Parameters	Exception
int getMinorDeviceClasses()	int—the major device class 0x0C—with a computer major device class, laptop minor device class 0x04—with a phone major device class, cellular minor device class	None	Throws: None

Class DiscoveryAgent

Description

The **DiscoveryAgent** class provides methods to perform device discovery (but not service discovery in leJOS NXJ). A local device must have only one **DiscoveryAgent** object. This object must be retrieved by a call to **getDiscoveryAgent()** on the **LocalDevice** object. The three service methods normally in the **DiscoveryAgent** class, such as **searchServices()**, are not included because the LEGO NXT brick only allows one service: Serial Port Profile (SPP). It would waste memory to implement the service methods considering that they are not really functional.

Device Discovery   There are two ways to discover devices. First, an application may use **startInquiry()** to start an inquiry to find devices in proximity to the local device. Discovered devices are returned via the **deviceDiscovered()** method of the interface

`DiscoveryListener`. The second way to discover devices is via the `retrieveDevices()` method. This method will return devices that have been discovered via a previous inquiry or devices that are classified as preknown. (Preknown devices are devices that are defined in the Bluetooth Control Center as devices this device frequently contacts.) The `retrieveDevices()` method does not perform an inquiry but provides a quick way to get a list of devices that may be in the area. Warning: If a device is found that has not yet been paired with the NXT brick, the name field of `RemoteDevice` will be blank. Make sure to pair your devices through the leJOS NXJ Bluetooth menu on your NXT.

cancelInquiry

Description: Removes the device from inquiry mode. An `inquiryCompleted()` event will occur with a type of `INQUIRY_TERMINATED` as a result of calling this method. After receiving this event, no further `deviceDiscovered()` events will occur as a result of this inquiry. This method will only cancel the inquiry if the `listener` provided is the listener that started the inquiry.

Prototype	Returns	Parameters	Exception
boolean cancelInquiry (DiscoveryListener listener)	boolean true— if the inquiry was canceled boolean false—if the inquiry was not canceled or if the inquiry was not started using listener	DiscoveryListener listener—the listener that is receiving inquiry events	Throws: NullPointerException—if listener is null

retrieveDevices

Description: Returns an array of Bluetooth devices that have either been found by the local device during previous inquiry requests or been specified as a preknown device depending on the argument. The list of previously found devices is maintained by the implementation of this API. (In other words, maintenance of the list of previously found devices is an implementation detail.) A device can be set as a preknown device in the Bluetooth Control Center.

Prototype	Returns	Parameters	Exception
RemoteDevices[] retrieveDevices(int option)	RemoteDevices[]— an array containing the Bluetooth devices that were previously found if option is CACHED; an array of devices that are preknown devices if option is PREKNOWN; null if no devices meet the criteria	int option— CACHED if previously found devices should be returned; PREKNOWN if preknown devices should be returned	Throws: IllegalArgument Exception—if option is not CACHED or PREKNOWN

startInquiry

Description: Places the device into inquiry mode. The length of the inquiry is implementation-dependent. This method will search for devices with the specified inquiry access code. Devices that responded to the inquiry are returned to the application via the method **deviceDiscovered()** of the interface **DiscoveryListener**. The **cancelInquiry()** method is called to stop the inquiry. Note: If a device is found that has not yet been paired with the NXT brick, the name field of RemoteDevice will be blank. Make sure to pair your devices through the leJOS NXJ Bluetooth menu on your NXT.

Prototype	Returns	Parameters	Exception
boolean startInquiry(int accessCode, DiscoveryListener listener) throws BluetoothStateException	boolean true— if the inquiry was started boolean false—if the inquiry was not started because the accessCode is not supported	int accessCode— the type of inquiry to complete DiscoveryListener listener—the event listener that will receive device discovery events	Throws: IllegalArgumentException—if the access code provided is not LIAC, GIAC, or in the range 0x9E8B00 to 0x9E8B3F NullPointerException—if listener is null BluetoothStateException—if the Bluetooth device does not allow an inquiry to be started owing to other operations that are being performed by the device

Class LocalDevice

Description

Singleton class representing a local NXT Bluetooth device. Most methods are standard, except you also can set the friendly name with this class.

getBluetoothAddress

Description: Returns the local Bluetooth address of the NXT brick.

Prototype	Returns	Parameters	Exception
String getBluetoothAddress()	String—the address	None	Throws: None

getDeviceClass

Description: The Bluetooth device class for the LEGO NXT brick. The LEGO Bluecore code can't retrieve this from the chip. Always returns hardcoded 0x3e0100 DeviceClass Untested whether this is the correct device class or not.

Prototype	Returns	Parameters	Exception
DeviceClass getDeviceClass()	DeviceClass—the device class	None	Throws: None

getDiscoveryAgent
Description: Returns the discovery agent for this device. Multiple calls to this method will return the same object. This method will never return null.

Prototype	Returns	Parameters	Exception
DiscoveryAgent getDiscoveryAgent()	DiscoveryAgent—the discovery agent for the local device	None	Throws: None

getDiscoverable
Description: Indicates whether the NXT brick is visible to other devices.

Prototype	Returns	Parameters	Exception
int getDiscoverable()	int 0—not discoverable, all others are discoverable	None	Throws: None

getFriendlyName
Description: Returns the friendly name of a Bluetooth device. Note: If you want to set the friendly name, it can be done through the **LCP** class or using the **NXJExplorer** program on your personal computer.

Prototype	Returns	Parameters	Exception
String getFriendlyName()	String—the friendly name	None	Throws: None

getLocalDevice
Description: Returns a static local NXT Bluetooth device.

Prototype	Returns	Parameters	Exception
static LocalDevice getLocalDevice()throws BluetoothStateException	LocalDevice—the local NXT Bluetooth device	None	Throws: BluetoothStateException

setFriendlyName
Description: Changes the friendly name of the NXT brick. Note: This method is not part of the standard JSR 82 API because not all Bluetooth devices can change their friendly name. This method does not work. Technically, this should be done through LCP, so USB also can change it.

Prototype	Returns	Parameters	Exception
boolean setFriendlyName(String name)	boolean true—if successful boolean false—if failed	String name—the new friendly	Throws: None

setDiscoverable
Description: Normally the mode values are found in `javax.bluetooth.DiscoveryAgent`. We don't have this yet in NXJ, so use 0 for invisible and any other value for visible.

Prototype	Returns	Parameters	Exception
boolean setDiscoverable(int mode) throws BluetoothStateException	boolean true—if successful boolean false—if failed	int mode—0 is invisible, all others visible	Throws: BluetoothStateException

getProperty
Description: Unimplemented! Returns `null` always. Returns various properties about the Bluetooth implementation, such as version, whether master-slave switch allowed, etc. Possibly use `Properties` class in implementation.

Prototype	Returns	Parameters	Exception
String getProperty(String property)	Null	String property	Throws: None

isPowerOn
Description: Power state of the Bluecore 4 chip in the NXT brick.

Prototype	Returns	Parameters	Exception
static boolean isPowerOn()	boolean—the power state	None	Throws: None

Class RemoteDevice

Description

Represents a remote Bluetooth device.

equals
Description: Determines if two RemoteDevices are equal. If they both have the same BT address, then they are considered equal. Overrides **equals()** in class `Object`.

Prototype	Returns	Parameters	Exception
boolean equals(Object obj)	boolean false—means it is not encrypted	Object obj	Throws: None

getBluetoothAddress
Description: Returns the local Bluetooth address of NXT brick.

Prototype	Returns	Parameters	Exception
`String getBluetoothAddress()`	`String`—the address	None	Throws: None

getDeviceAddr
Description: Returns the address of `RemoteDevice`.

Prototype	Returns	Parameters	Exception
`static String getDeviceAddr()`	`String`—address	None	Throws: None

getDeviceClass
Description: The Bluetooth device class for the LEGO NXT brick. The LEGO Bluecore code can't retrieve this from the chip. Always returns hardcoded `0x3e0100 DeviceClass Untested` whether this is the correct device class or not.

Prototype	Returns	Parameters	Exception
`int getDeviceClass()`	int	None	Throws: None

getFriendlyName
Description: Returns the friendly name of the **RemoteClass**.

Prototype	Returns	Parameters	Exception
`String getFriendlyName(boolean alwaysAsk)`	`String`—the friendly name	`boolean true`— causes the method to contact the remote device for the name `boolean false`—it will use the known name	Throws: None

getRemoteDevice

Prototype	Returns	Parameters	Exception
`static RemoteDevice getRemoteDevice(Connection conn) throws IOException`	`RemoteDevice`—a remote Bluetooth device	`Connection conn`— communication connection	Throws: `IOException`

isAuthenticated
Description: Indicates if the remote device has been properly paired (authenticated).

Prototype	Returns	Parameters	Exception
`boolean isAuthenticated()`	`boolean true`—if the device has been paired	None	Throws: None

isEncrypted
Description: Always returns false because NXT BC4 chip doesn't encrypt.

Prototype	Returns	Parameters	Exception
`boolean isEncrypted()`	`boolean false`—means it is not encrypted	None	Throws: None

RemoteDevice
Constructor: The standard JSR 82 method for obtaining a **RemoteDevice** uses a string rather than `byte[]`. Protected, so shouldn't matter.

Prototype	Returns	Parameters	Exception
`RemoteDevice(String addr)`	None	`String addr`	Throws: None

RemoteDevice
Constructor: Constructs a **RemoteDevice** object with a name, device address and the device class.

Prototype	Returns	Parameters	Exception
`RemoteDevice(String name, String deviceAddr, int devclass)`	None	`String name` `String deviceAddr` `int devclass`	Throws: None

setDeviceAddr
Description: Sets the device address to `deviceAddr`.

Prototype	Returns	Parameters	Exception
`void setDeviceAddr(String deviceAddr)`	`void`	`String deviceAddr`	Throws: None

leJOS Bluetooth API

Class NXTCommDevice

Description

Base class for NXT communications devices. Provides a common address/name plus utility functions.

addressToString
Description: Helper method to convert address byte array to String.

Prototype	Returns	Parameters	Exception
static String addressToString(byte[] addr)	String—string representation of Bluetooth address	byte[] addr—a byte array of bytes containing the address	Throws: None

getAddress
Description: Returns the current USB serial number.

Prototype	Returns	Parameters	Exception
static String getAddress()	String—the current serial number	None	Throws: None

getName
Description: Returns the current USB devName.

Prototype	Returns	Parameters	Exception
static String getName()	String—the devName	None	Throws: None

isAddress
Description: Determines if a string contains a Bluetooth-style address.

Prototype	Returns	Parameters	Exception
static boolean isAddress(String s)	boolean—true—if the string is an address	String s—String to test	Throws: None

loadSettings
Description: Loads the current system settings associated with this class. Called automatically to initialize the class. May be called if it is required to reload any settings.

Prototype	Returns	Parameters	Exception
static void loadSettings()	void	None	Throws: None

nameToString
Description: Returns a string version of a device **devName** held as a byte array.

Prototype	Returns	Parameters	Exception
static String nameToString(byte[] name)	String—string version of devName	byte[] name	Throws: None

setAddress
Description: Sets the USB serial number.

Prototype	Returns	Parameters	Exception
static void setAddress(String sn)	void	String sn	Throws: None

setName
Description: Set the USB **devName**.

Prototype	Returns	Parameters	Exception
static void setName(String name)	void	String name—the name	Throws: None

stringToAddress
Description: Converts a string version of a Bluetooth address into a byte-array address.

Prototype	Returns	Parameters	Exception
static byte[] stringToAddress(String strAddress)	byte[]—a byte-array version of the address	String strAddress—the string version of the address	Throws: None

stringToName
Description: Converts the string version of a **devName** into a byte array.

Prototype	Returns	Parameters	Exception
static byte[] stringToName(String strName)	byte[]—a byte-array version of the name	String strName—string version of the devName	Throws: None

Class `Bluetooth`

Description
Provides Bluetooth communications. Allows inbound and outbound connections. Provides access to device registration.

btEnable
Description: None.

Prototype	Returns	Parameters	Exception
static void btEnable()	void	None	Throws: None

btDisable
Description: None.

Prototype	Returns	Parameters	Exception
static void btDisable()	void	None	Throws: None

btGetBC4CmdMode
Description: Low-level method to get the BC4 chip mode.

Prototype	Returns	Parameters	Exception
static int btGetBC4CmdMode()	int—the current mode	None	Throws: None

btRead
Description: Low-level method to read BT data.

Prototype	Returns	Parameters	Exception
static int btRead(byte[] buf,int off, int len)	int—number of bytes actually read	byte[]buf—the buffer to read data into int off—the offset at which to start the transfer int len—the number of bytes to read	Throws: None

btReceive
Description: Low-level method to receive BT replies or data.

Prototype	Returns	Parameters	Exception
static void btReceive(byte[]buf)	void	byte[]buf—the buffer to receive data in	Throws: None

btPending

Description: Low-level method to access the Bluetooth interface. Bitwise values returned.

Prototype	Returns	Parameters	Exception
static int btPending()	int—0 means no data pending, 0x1 means input pending, and 0x2 means output pending.	None	Throws: None

btSetArmCmdMode

Description: Low-level method to switch BC4 chip between command and data (stream) mode.

Prototype	Returns	Parameters	Exception
static void btSetArmCmdMode(int mode)	void	int mode—0 is data mode, 1 is command mode	Throws: None

btSetCmdMode

Description: Sets the BC4 mode and waits for that mode to be confirmed by the chip.

Prototype	Returns	Parameters	Exception
static void btSetCmdMode(int mode)	void	int mode—the requested mode, 1 is the Command mode, 0 is the Stream mode	Throws: None

btSetResetHigh

Description: Low-level method to take the BC4 reset line high.

Prototype	Returns	Parameters	Exception
static void btSetResetHigh()	void	None	Throws: None

btSetResetLow

Description: Low-level method to take the BC4 reset line low.

Prototype	Returns	Parameters	Exception
static void btSetResetLow()	void	None	Throws: None

btStartADConverter

Description: Low-level method to access the Bluetooth interface.

Prototype	Returns	Parameters	Exception
static void btStartADConverter()	void	None	Throws: None

btSend
Description: Low-level method to send a BT command or data.

Prototype	Returns	Parameters	Exception
`static void btSend(byte[] buf,int len)`	void	byte[]buf—the buffer to send int len—the number of bytes to send	Throws: None

btWrite
Description: Low-level method to write BT data.

Prototype	Returns	Parameters	Exception
`static int btWrite(byte[] buf,int off, int len)`	int—number of bytes actually written	byte[]buf—the buffer to send int off—the offset to start the write from int len—the number of bytes to send	Throws: None

addDevice
Description: Adds device to known devices.

Prototype	Returns	Parameters	Exception
`static boolean addDevice(RemoteDevice d)`	boolen true—if add is successful	RemoteDevice d—remote device	Throws: None

cancelInquiry
Description: Cancels a Bluetooth inquiry process that has been started using `startInquire`.

Prototype	Returns	Parameters	Exception
`static boolean cancelInquiry()`	boolean—true if the request is canceled, false if there is an error	None	Throws: None

closeConnection
Description: Closes an open connection.

Prototype	Returns	Parameters	Exception
`static int closeConnection(byte handle)`	int handle—the handle for the connection	byte handle—the status 0 is success	Throws: None

closePort
Description: Closes the port to disallow incoming connections.

Prototype	Returns	Parameters	Exception
`static byte[] closePort()`	`byte[]`—an array of 2 bytes: `success, ps_success`	None	Throws: None

connect
Description: Connects to a remote device. Uses the current default pin.

Prototype	Returns	Parameters	Exception
`static BTConnection connect(RemoteDevice remoteDevice)`	`BTConnection`—the Bluetooth connection object or `null`	`RemoteDevice remoteDevice`— remote device	Throws: None

connect
Description: Connect to the specified device, either by name or by address.

Prototype	Returns	Parameters	Exception
`static BTConnection connect(String target, int mode, byte[] pin)`	`BTConnection`—the Bluetooth connection object or `null`	`String target`— String name or address `int mode`— the I/O mode for this connection: `NXTConnection.RAW`, `.LCP`, or `.PACKET` `byte[] pin`—the pin to use	Throws: None

connect
Description: Connects to the specified device, either by name or by address.

Prototype	Returns	Parameters	Exception
`static BTConnection connect(String target, int mode)`	`BTConnection`—the Bluetooth connection object or `null`	`String target`— String name or address `int mode`—the I/O mode for this connection: `NXTConnection.RAW`, `.LCP`, or `.PACKET`	Throws: None

getConnectionStatus
Description: Gets the status of all connections.

Prototype	Returns	Parameters	Exception
static byte[] getConnectionStatus()	byte[]—byte array of status for each handle	None	Throws: None

getConnector
Description: Provides access to the singleton `Connection` object. This object can be used to create new connections.

Prototype	Returns	Parameters	Exception
static NXTConnector getConnector()	NXTConnector—the Connection object	None	Throws: None

getFriendlyName
Description: Gets the friendly name of the local device.

Prototype	Returns	Parameters	Exception
static String getFriendlyName()	String—the friendly name	None	Throws: None

getKnownDevice
Description: Gets a device of the BC4 chip's internal list of known devices (those which have been paired before) into the `RemoteDevice` object.

Prototype	Returns	Parameters	Exception
static RemoteDevice getKnownDevice(String fName)	RemoteDevice—object or null if not found	String fName—friendly name of device	Throws: None

getKnownDevicesList
Description: The internal chip has a list of already paired devices. This method returns a vector list that contains all the known devices on the list. These need not be reachable. To connect to a "not known" device, you should use the inquiry process. The pairing process also can be done with the original LEGO firmware. The list of known devices will not get lost when installing the leJOS firmware.

Prototype	Returns	Parameters	Exception
static Vector getKnownDevicesList()	Vector—vector with list of known devices	None	Throws: None

getLocalAddress
Description: Gets the Bluetooth address of the local device.

Prototype	Returns	Parameters	Exception
`static String getLocalAddress()`	String—the local address	None	Throws: None

getOperatingMode
Description: Adds device to known devices.

Prototype	Returns	Parameters	Exception
`static int getOperatingMode()`	int—0 = stream-breaking mode, 1 = don't break stream mode < 0 error	None	Throws: None

getPin
Description: Returns the pin to be used for pairing/connecting to the system.

Prototype	Returns	Parameters	Exception
`static byte[] getPin()`	byte[]int—the current pin code	None	Throws: None

getPortOpen
Description: Gets the port open status, i.e., whether connections are being accepted.

Prototype	Returns	Parameters	Exception
`static int getPortOpen()`	int—1 if the port is open, 0 otherwise.	None	Throws: None

getPower
Description: Sets the power to the module.

Prototype	Returns	Parameters	Exception
`static boolean getPower()`	boolean—power on or off	None	Throws: None

getResetCount
Description: Returns the reset count.

Prototype	Returns	Parameters	Exception
`static int getResetCount()`	int—reset count	None	Throws: None

getSignalStrength
Description: Gets the Bluetooth signal strength (link quality). Higher values mean stronger signal.

Prototype	Returns	Parameters	Exception
`static int getSignalStrength(byte handle)`	int—link quality value 0 to 255	`byte handle`—the handle/channel of the connection	Throws: None

getStatus
Description: Gets the persistent status value from the BC4 chip.

Prototype	Returns	Parameters	Exception
`static int getStatus()`	byte[]—the byte value	None	Throws: None

getVersion
Description: Gets the major and minor version of the BlueCore code.

Prototype	Returns	Parameters	Exception
`static byte[] getVersion()`	byte[]—an array of 2 bytes: major version, minor version	None	Throws: None

getVisibility
Description: Gets the visibility (discoverable) status of the device.

Prototype	Returns	Parameters	Exception
`static int getVisibility()`	int—1 = visible, 0 = invisible	None	Throws: None

Inquire
Description: Starts a Bluetooth inquiry process.

Prototype	Returns	Parameters	Exception
`static Vector inquire(int maxDevices, int timeout, byte[] cod)`	Vector—a vector of all the devices found	`int maxDevices`—the maximum number of devices to discover `int timeout`—the timeout value in units of 1.28 seconds `byte [] cod`—the class of device to look for	Throws: None

inquireNotify

Description: Starts a Bluetooth inquiry process and notifies listener of each device found. This method is used primarily by `DiscoveryAgent.startInquiry()` and is run in a separate thread.

Prototype	Returns	Parameters	Exception
`static void inquireNotify(int maxDevices, int timeout, DiscoveryListener listy)`	`void`	`int maxDevices`—the maximum number of devices to discover `int timeout`—the timeout value in units of 1.28 seconds `DiscoverListener listy`—the listener to notify	Throws: None

loadSettings

Description: None.

Prototype	Returns	Parameters	Exception
`static void loadSettings()`	`void`	None	Throws: None

lookupName

Description: Looks up the name of a device using its address.

Prototype	Returns	Parameters	Exception
`static String lookupName(String addr)`	`String`—friendly name of device	`String addr`—device address	Throws: None

openPort

Description: Opens the port to allow incoming connections.

Prototype	Returns	Parameters	Exception
`static byte[] openPort()`	`byte[]`—an array of 3 bytes: success, handle, ps_success	None	Throws: None

removeDevice

Description: Removes device from known devices.

Prototype	Returns	Parameters	Exception
`static boolean removeDevice(RemoveDevice d)`	`boolen true`—if remove is successful	`RemoveDevice d`—remove device	Throws: None

reset
Description: Forces a reset of the Bluetooth module. Note: After this call, power will be on. Any existing connections will be closed. Any listening threads will be aborted.

Prototype	Returns	Parameters	Exception
static void reset()	void	None	Throws: None

setFactorySettings
Description: Resets the settings of the BC4 chip to the factory defaults. The NXT should be restarted after this.

Prototype	Returns	Parameters	Exception
static int setFactorySettings()	int—0 if okay, <0 if error	None	Throws: None

setFriendlyName
Description: Sets the name of the local device.

Prototype	Returns	Parameters	Exception
static boolean setFriendlyName(String strName)	Boolean—true if okay, false if there is an error	String strName—the friendly name for the device	Throws: None

setOperatingMode
Description: Sets the operating mode.

Prototype	Returns	Parameters	Exception
static int setOperatingMode(byte mode)	int—<0 error	byte mode—0 = stream breaking, 1 = don't break stream	Throws: None

setPin
Description: Sets the pin to be used for pairing/connecting to the system.

Prototype	Returns	Parameters	Exception
static void setPin(byte[] newPin)	void	byte[ ] newPin—the new pin code	Throws: None

setPower
Description: Sets the power to the module.

Prototype	Returns	Parameters	Exception
static void setPower(boolean on)	void	boolean on—power on or off	Throws: None

setStatus
Description: Sets the persistent status byte for the BC4 chip.

Prototype	Returns	Parameters	Exception
`static int setStatus(int status)`	int—<0 error	`int status`—the byte status value	Throws: None

setVisibility
Description: Sets Bluetooth visibility (discoverable) on or off for the local device.

Prototype	Returns	Parameters	Exception
`static int setVisibility(byte visible)`	int—<0 error	`byte visible`— true to set visibility on, `false` to set it off	Throws: None

waitForConnection
Description: Low-level method to access the Bluetooth interface. Bitwise values returned.

Prototype	Returns	Parameters	Exception
`static BTConnection waitForConnection(int timeout,int mode, byte[] pin)`	BTConnection—a Bluetooth connection	`int timeout`—time in microseconds to wait for connection, 0 is wait forever `int mode`—the I/O mode to be used for this connection. `NXTConnection.RAW`, `.LCP`, or `.PACKET` `byte[] pin`—the pin to use, `null` use current default	Throws: None

waitForConnection
Description: Uses the current default PIN.

Prototype	Returns	Parameters	Exception
`static BTConnection waitForConnection()`	BTConnection—the Bluetooth connection	None	Throws: None

waitForConnection
Description: Uses the current default PIN.

Prototype	Returns	Parameters	Exception
`static BTConnection waitForConnection(int timeout,int mode)`	BTConnection—the Bluetooth connection	`int timeout`—time in microseconds to wait for connection, 0 is wait forever `int mode`—the I/O mode to be used for this connection: `NXTConnection.RAW`, `.LCP`, or `.PACKET`	Throws: None

Class NXTConnection

Description
Generic lejosnxt connection class. Provides access to standard read/write methods. This code supports both asynchronous (used for Bluetooth and RS-485 connections) and synchronous (used for USB) input-output (I/O) operations for the actual reading and writing of the low-level buffers.

available
Description: Indicates the number of bytes available to be read. Supports both PACKET mode and Stream connections.

Prototype	Returns	Parameters	Exception
int available(int what)	int—number of bytes available	int what—0 (all modes) returns the number of bytes that can be read without blocking; 1 (packet mode) returns the number of bytes still to be read from the current packet; 2 (packet mode) returns the length of the current packet	Throws: None

available
Description: Convenience method that calls available(0).

Prototype	Returns	Parameters	Exception
int available()	int—number of bytes available	None	Throws: None

close
Description: Closes the connection. Flushes any pending output. Informs the remote side that the connection is now closed and frees resources.

Prototype	Returns	Parameters	Exception
static void close()	void	None	Throws: None

openDataInputStream
Description: Returns the DataInputStream for this connect.

Prototype	Returns	Parameters	Exception
DataOutputStream openDataOutputStream()	DataOutputStream—the data output stream	None	Throws: None

read
Description: Performs a blocking read on the connection.

Prototype	Returns	Parameters	Exception
`int read(byte[] data,int len)`	int—actual number of bytes read, return <0 for error.	`byte[] data`—byte array to store results `int len`—max. number of bytes to read	Throws: None

readPacket
Description: Reads a packet from the stream. Does not block and for small packets (< [package access scope variable] `bufSz`), does not return a partial packet.

Prototype	Returns	Parameters	Exception
`static int readPacket(byte[] buf, int len)`	int—>0 number of bytes read; other values, see read	`byte[]buf`—buffer to read data into `int len`—number of bytes to read	Throws: None

sendPacket
Description: Sends a data packet. Must be in data mode.

Prototype	Returns	Parameters	Exception
`static int sendPacket(byte[] buf, int bufLen)`	void	`byte[]buf`—the data to send `int bufLen`—the number of bytes to send	Throws: None

setIOMode
Description: Sets operating mode. Controls the packet/stream mode of this channel. For PACKET mode, it defines the header size to be used.

Prototype	Returns	Parameters	Exception
`void setIOMode(int mode)`	void	`int mode`—I/0 mode to be used for this connection: RAW, LCP, or PACKET	Throws: None

write
Description: Performs a blocking write on the connection.

Prototype	Returns	Parameters	Exception
`int write(byte[] data, int len)`	int—actual number of bytes written, return <0 for error	`byte[] data`—byte array to be written `int len`—number of bytes to write	Throws: None

write

Description: Attempts to write bytes to the connection. Optionally waits if it is not possible to write at the moment. Supports both **PACKET** and **RAW** write operations. If in **PACKET** mode, a set of header bytes indicating the size of the packet will be sent ahead of the data.

Prototype	Returns	Parameters	Exception
`int write(byte[] data, int len,boolean wait)`	`int—>0`: number of bytes written 0: Request would have blocked (and wait was false) −1: An error occurred −2: Data have been lost	`byte[] data`—byte array to be written `int len`—number of bytes to write `boolean wait—true` if the call should block until all the data have been sent	Throws: `None`

Class `BTConnection`

Description

Provides a Bluetooth connection. Supports both packetized, raw, and stream-based communication. Blocking and nonblocking I/O. Note: Because of the limited buffer space and the way that several connections have to share the interface with the Bluetooth device, data may be lost. This will happen if a switch into command mode is required when data are arriving from the remote connection that cannot be placed into the input buffer. Every attempt is made to avoid this, but it can happen. Application programs can help to avoid this problem by (1) using just a single Bluetooth connection, (2) using Bluetooth commands while data transfers are in progress, (3) performing application-level flow control to avoid more than 256 bytes of data being sent from the remote side at any one time, and (4) reading any pending data as soon as possible. If data are lost, then calls to read and write will return −2 to indicate the problem. If using packet mode, then the input stream can be resynchronized by issuing a read to discard the partial packet that may be in the input buffer. When operating in **RAW** mode, bytes are read/written as is. This mode is useful for talking to leJOS/LEGO devices. When operating in **PACKET** mode, the standard LEGO 2-byte header is added to each packet (and is expected to be present on each incoming packet). Use this mode when talking to other leJOS/LEGO devices.

closeStream

Description: Closes the stream for this connection. This suspends the connection and switches the BC4 chip to command mode.

Prototype	Returns	Parameters	Exception
`void closeStream()`	`void`	None	Throws: `None`

getSignalStrength

Description: Gets the signal strength of this connection. This necessitates closing and reopening the data stream.

Prototype	Returns	Parameters	Exception
int getSignalStrength()	int—a value from 0 to 255	None	Throws: None

setActiveMode

Description: Sets the channel-switching mode. Allows control of when we will switch to this channel. By default, we will switch to this channel to check for input. However if **AM_OUTPUT** is set, we only switch if we have output waiting to be sent.

Prototype	Returns	Parameters	Exception
void setActiveMode(int mode)	void	int mode—the switch control mode	Throws: None

Appendix B

Bluetooth Robotic-Oriented Network (BRON) Team Members

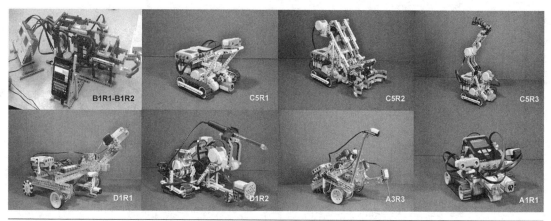

FIGURE B-1 Grid of the LEGO/TETRIX robotics systems BRON team members.

BRON Cube Solver Team

The Rubik's Cube solver team is a collective of two NXT microcontrollers and a Android-based smart phone. These are components of one robot within a fixed frame whose purpose is to solve the infamous Rubik's Cube. Communication and coordination are directed at physically manipulating the cube and analyzing which colors are where and how to get them where they need to be. An Android phone takes pictures of all sides of the cube. A generated solution is a set of moves to solve the cube. These moves are sent to two NXT bricks, each of which controls the end effectors. The two end effectors manipulate the cube, which solves the puzzle.

TABLE B-1 The Cube Solver

Team B Members	Microcontroller	End Effectors	Mobility Type	Display	Sensors/ Motors	I/O Port	Ranges	Other
Flippy (B1R1)	32-bit ARM	Flipper	N/A	100- × 64-pixel liquid-crystal display (LCD) screen	Flipper (linear actuator)	A	–120–0 degrees	N/A
		Gripper			Gripper (linear actuator)	B	–45–0 degrees	

TABLE B-1 The Cube Solver

Team B Members	Microcontroller	End Effectors	Mobility Type	Display	Sensors/ Motors	I/O Port	Ranges	Other
Twisty (B1R2)	32-bit ARM	Upper: Twister (rotating platform, oriented upside-down)	N/A	100- × 64-pixel LCD screen	Upper twister	A	0–360 degrees	N/A
		Lower: Twister (rotating platform)			Lower twister	B	0–360 degrees	
		Aligner: Two fingers moving in parallel			Aligner	C	-2300–0 degrees	

BRON Convoy Team

This is a convoy of three NXT-based robots with designations C5R1, C5R2, and C5R1. The team leader (C5R1) in the convoy initiates communication with the other two team members (C5R2 and C5R3). All communication originates from the team leader (C5R1). Each team member (C5R2 and C5R3) listens for a command to be sent from C5R1. On receipt of that command, each executes a program. If they did not receive the command, the program that is executed is a stop program. If they received the command but for some reason did not execute the program, a reply is sent back to the team leader to stop the convoy.

TABLE B-2 Team Member C5R1

Team C Member	Microcontroller	End Effectors	Mobility Type	Display	Sensors/ Motors	I/O Port	Ranges	Other
C5R1	32-bit ARM	N/A	Tractor	100- × 64-pixel LCD	Ultrasonic	4	0–255 cm	N/A
					Color	3	0–6	
					Servo left wheel	C	0–360 degrees	
					Servo right wheel	B	0–360 degrees	

TABLE B-3 Team Member C5R2

Team C Member	Microcontroller	End Effectors	Mobility Type	Display	Sensors/ Motors	I/O Port	Ranges	Other
C5R2	32-bit ARM	Gripper	Tractor	100- × 64-pixel LCD	Ultrasonic	4	0–255 cm	N/A
					Color	3	0–6	
					Touch	1	0–1	
					Servo left wheel	C	0–360 degrees	
					Servo right wheel	B	0–360 degrees	
					Servo gripper	A	0–360 degrees	

TABLE B-4 Team Member C5R3

Team C Member	Microcontroller	End Effectors	Mobility Type	Display	Sensors/ Motors	I/O Port	Ranges	Other
C5R3	32-bit ARM	N/A	Tractor	100- × 64-pixel LCD	Ultrasonic	4	0–255 cm	N/A
					Servo left wheel	C	0–360 degrees	
					Servo right wheel	B	0–360 degrees	

BRON Crime Scene Investigation (CSI) Team

In this project, the sentry robot (A1R1) investigates, secures, and reports any anomalies that it encounters in a warehouse. The sentry robot makes its rounds of the warehouse randomly using radio frequency identification (RFID) to identify objects. The sentry robot encounters an unknown container, performs an immediate analysis, and sends for more members of the CSI team. Robot D1R1 uses its robotic arm to secure the container. The CSI members perform a more in-depth analysis and determine that the container holds an unknown liquid. If the substance is harmless, a report is made using an Android smart phone. If the substance is harmful, an emergency call is placed to several agencies of concern.

TABLE B-5   Team Member D1R1

Team D Member	Microcontroller	End Effectors	Mobility Type	Display	Sensors/ Motors	I/O Port	Ranges	Other
D1R1	32-bit ARM7, 64-kB RAM	Arm and gripper	Four wheels	100- × 64-pixel LCD screen	LEGO touch	3	1–0	Bluetooth, USB
					HiTechnic color	2	16 colors	
					LEGO ultrasonic	4	0–255 cm	
					DC motor controller	1	360 degrees	
					Left wheel DC motor	DC controller 1–, 1+	360 degrees	
					Right wheel DC motor	DC controller 2–, 2+	360 degrees	
					Left wheel encoder	DC motor controller Enc. 1	360 degrees	
					Right wheel encoder	DC motor controller Enc. 2	360 degrees	
					Arm servo	Servo controller, channel 1	90 degrees	
					Gripper servo	Servo controller, channel 2	90 degrees	

TABLE B-6   Team Member D1R2

Team D Member	Microcontroller	End Effectors	Mobility Type	Display	Sensors/ Motors	I/O Port	Ranges	Other
D1R2	32-bit ARM7, 64-kB RAM	Probe and touch sensor extension	Two tractor wheels	100- × 64-pixel LCD screen	LEGO touch	1	1–0	Bluetooth, USB
					Vernier pH sensor adapter	3	pH 1–14	
					Probe servo	B	90 degrees	
					Right tractor wheel servo	C	360 degrees	
					Left tractor wheel servo	A	360 degrees	

TABLE B-7   Team Member A3R3

Team A Member	Micro-controller	End Effectors	Mobility Type	Display	Sensors/ Motors	I/O Port	Ranges	Other
A3R3	32-bit ARM7, 64-kB RAM	Arm with probe effector	Three wheels	100- × 64-pixel LCD screen	LEGO temperature probe	2	–20– 120°C, –4–248°F	Bluetooth, USB
					Ultrasonic	4	0–255 cm	
					HiTechnic compass	3	0	
					DC motor right wheel	DC motor controller motor 1–, 1+	360 degrees	
					DC motor left wheel	DC motor controller motor 2–, 2+	360 degrees	
					Right wheel encoder	DC motor controller Enc. 1	N/A	
					Left wheel encoder	DC motor controller Enc. 2	N/A	
					DC controller	1	N/A	
					Right tractor wheel servo	C	360 degrees	
					Left tractor wheel servo	A	360 degrees	
					Servo motor for arm	Servo controller. channel 2	180 degrees	

TABLE B-8 Team Member A1R1

Team A Member	Microcontroller	End Effectors	Mobility Type	Display	Sensors/ Motors	I/O Port	Ranges	Other
A1R1	32-bit ARM7, 64-kB RAM	N/A	Tractor wheel	100- × 64-pixel LCD screen	Ultrasonic	4	0–255 cm	Bluetooth, USB
					CODATEX RFID	1	5–12 bytes ID	
					LEGO color	3	0–6 colors	
					Servo right wheel	B	0–360 degrees	
					Servo left wheel	C	0–360 degrees	
					Servo center	B	0–360 degrees	

Index